SIMULATING THE COSMOS

● UNIVERSE

SERIES EDITOR: James Geach

Titles in the Series
Simulating the Cosmos: Why the Universe Looks the Way It Does ROMEEL DAVÉ
Twenty Worlds: The Extraordinary Story of Planets Around Other Stars NIALL DEACON

Forthcoming
Telescopes / Observing the Universe SARAH KENDREW

SIMULATING

THE

COSMOS

*Why the Universe
Looks the Way It Does*

ROMEEL DAVÉ

REAKTION BOOKS

To my wife and daughters, the shining stars in my life

Published by
REAKTION BOOKS LTD
Unit 32, Waterside
44–48 Wharf Road
London N1 7UX, UK
www.reaktionbooks.co.uk

First published 2023
Copyright © Romeel Davé 2023

Printed and bound in India by Replika Press Pvt. Ltd

A catalogue record for this book is available from the British Library

ISBN 978 1 78914 714 8

CONTENTS

INTRODUCTION

Hi! My name is Romeel, and I'm a numerical cosmologist. We numerical cosmologists are preoccupied with answering one big question. It's a question that is incredibly simple to ask, but deceptively tricky to answer, a bit like when your partner asks you, 'Where have you been all evening?' But this question is even harder, even bigger and even more profoundly dangerous:

Why does the Universe look the way it does?

This might be the biggest question of all. It is a question whose answer underpins the entire story of our existence, asked by humans throughout the millennia of our time as a sentient species, spawning both inexhaustible wonder as well as deep existential crises. It is a question that has been used to consolidate power, to inspire devotion and to start wars. It is the question that lies at the very inception of the creation story of human beings.

This big question has been stated and debated in many different ways. What was the origin of the Universe around us? Is the Universe eternal or did it have a beginning? If it had a beginning, when was that? And if not, exactly how is that possible? Did the Universe always look the same or has it changed over time? Could it potentially have looked very different or did it have no choice but to look the way it does? What does the Universe's future hold?

And of course the most provocative question of them all: do we live in a simulation?

We numerical cosmologists are trying to answer these questions by using computers. The way we figure it is, why should we bother doing the thinking if we can get the computer to do the thinking for us? This seems like a fair division of labour. The annoying reality is that numerical cosmologists still have to tell the computer what to think in very precise terms, as computers tend to be a bit pernickety. This turns out to be harder than it sounds.

Yet for all this, partly through ingenuity, partly through stupendous advances in computing power and partly through dumb luck, we have made amazing progress. We can now generate a reasonable facsimile of the real Universe on a computer, using the laws of physics as we understand them today. (Note: throughout this book I will use Universe with a capital 'U' to denote the real Universe and a little 'u' to denote simulated universes.) Maybe we can't get down to such details as planets and people and penguins, but these fake universes look surprisingly realistic on the scales of galaxies and above. This success has only been achieved in the last decade or so, and attests to the mind-boggling rate of progress in numerical cosmology.

It is a remarkable triumph of modern physics that we now have a plausible story for how humans got here, starting from the Big Bang all the way to the present-day Universe – even if some T's and C's still apply. To do so, we have had to pull together every field of modern physics, from quantum mechanics and atomic physics to Einstein's theory of relativity, and weave them together in just the right way. In what is either a vast cosmic conspiracy or a rare glimpse of the underlying truth, the pieces all basically fit. Cosmology and galaxy formation can take us from the Big Bang to our Milky Way galaxy, at which point star and planet formation tells us how the Sun and the Earth came to be formed. From there, we move on over to the chemists and biologists who seem to have a sensible story for the journey that began at the formation of our planet and advanced to the development of life, which ultimately evolved into humans. We have, for the first time in human history, a scientifically based verifiable Creation Story, taking us all the way back to the birth of our Universe – in no small part, thanks to computers.

In this book, I'll give you a flavour of what it takes to build a universe on a computer. Our framework is built on the now well-established concordance cosmological model, from which we will see how to fit a Universe into a computer and bake it for almost 14 billion years in our silicon oven. Along the way I'll highlight some of the new discoveries revealed by the first generation of simulation, such as the emergence of the cosmic web. Then we will learn about galaxies, the faint blobs of light that astronomers use to map out the Universe's history over most of cosmic time. I'll introduce you to the beautiful diversity of galaxies, which was recognized in the 1920s but is still not fully understood to this day, even with spectacular observations from the latest and greatest telescope facilities. I'll lead you through how recent simulations have revealed that galaxies are not 'island universes', as presciently postulated by Immanuel Kant in the eighteenth century, but rather live and grow within a vibrant ecosystem mostly hidden from even our most sophisticated telescopes. We will finish at the bleeding edge of numerical cosmology, discussing some of the biggest outstanding questions about why the Universe looks the way it does.

By the end of this book, if it has done exactly what it says on the tin, you'll have a good idea of the progress we have made towards understanding our scientifically based cosmic origins story, and how supercomputer simulations have played a key role in this. But you'll also see how far we have yet to go, and how many pieces of the puzzle remain shrouded in mystery. It's a fascinating time to be a numerical cosmologist – one might even call it a golden age. If nothing else I hope this book will give you a glimpse into the excitement, intrigue and wonder that abounds in our corner of science. Enjoy the journey!

WHAT A LONG STRANGE TRIP IT'S BEEN

I got my first computer when I was eleven years old. It was an Atari 400, and it cost about $600 (£250 in 1980), which at that time was a small fortune for our family. It hooked up to our TV through an RCA jack. The hard drive was a cassette player, and you could run one 'app' at a time by sticking a corresponding cartridge into a bulky slot at the top of the machine.

Like any right-thinking eleven-year-old boy, I wanted to play video games. My father said, 'Sure! All you have to do is learn the Atari BASIC programming language, and program your own!' He handed me a BASIC cartridge and a manual. Not exactly what I had in mind.

So programming it was. It helped that my father was an electrical engineer who knew how computers worked. But to say it was confusing and frustrating would be a massive understatement. I had never been so annoyed by anything in my life. The stubborn Atari just kept doing what I *told* it to do, instead of what I so obviously *wanted* it to do!

Programming is an entirely different mindset. I knew some algebra by that age, at least enough to know that the statement $n = n + 1$ could not possibly ever be true. I had to retrain my mind to think of computer programs as a recipe for the computer to follow. The equation $n = n+1$ is actually shorthand for the following procedure: go into the system memory, grab the value of n, add 1 to it and store it back in the same place in the computer's memory. Once I had internalized the idea of computer programs as recipes, it started to make a lot more sense. But it did not come easily.

I managed to write a simple game – a single alien spaceship scrolling across the top of the computer screen which you had to shoot with a missile from another spaceship that you could move across the bottom. The game was certainly nothing elaborate, but I was very proud of having written it myself. Sure, my friend had a fancier Atari 800 with a proper *Space Invaders* cartridge, but that didn't stop me from playing my game for hours on end, occasionally stopping to add some new twist into the code. In a way, it was fun to have control over my personal gaming universe.

My love/hate relationship with computers lasts to this very day. Although it is my job to work on computers all day, I can't say that I have any particular affinity for them. A computer is like the most pedantic and stubborn mule that has ever lived, who will refuse to budge or will wander off in a completely wrong direction if any of the instructions are in any way incomplete, lack specificity or are potentially contradictory. Sure, computers can do computations far faster than humans, particularly repetitive ones that would bore us to tears, but getting computers to do exactly the calculations you want . . . that's another story.

Fast-forward to my years as a graduate student in physics at the California Institute of Technology (Caltech) circa 1990. While chasing my dream of being a string theorist like some real-life Sheldon Cooper from *The Big Bang Theory*, I stumbled onto a field called cosmology. Cosmology aims to find the answer to the question: what is the origin and evolution of the Universe? I love big questions, and you don't get a much bigger question than that! I came to realize that there were people doing computer simulations of the universe. Naive wannabe string theorist that I was at the time, I had no idea that this was even a subject to specialize in.

To me, cosmology seemed like the perfect marriage of computing technology and fundamental physics. Here was an opportunity to leverage the crazy pace of advancement in computing technology to do science in a totally new way. For me personally, it was a chance to bring together two areas, physics and programming, I was slightly more knowledgeable than your average bear. I was intrigued by the whole idea. How exactly does one go about simulating an entire universe? How accurate and representative would it be? What are the challenges involved? If the computer simply follows the instructions that we feed it, how can we ever learn anything new?

I transferred to the University of California, Santa Cruz, whose faculty had some of the leading experts in cosmological simulations at the time, including a young faculty member named Lars Hernquist (now the Mallinckrodt Professor at Harvard University), who agreed to be my PhD advisor. By the end of my doctorate, I was a fully fledged numerical cosmologist, ready to do my part to help put the universe into a computer. And so began my journey into the exciting and dynamic realm of simulating the cosmos.

HUMANS AND THE SKY

One of the gifts of being a cosmologist is the sense of being part of a larger effort of humankind, one that spans the aeons of our species' existence. In the halcyon days before light pollution and industrial fog, for millennia humans stared into the night sky and asked the very same question that numerical cosmologists ask today: why does the universe look the way it does?

Perhaps the greatest trait separating human beings from other animals is our sense of wonder. Faced with such unanswerable yet fundamental questions, humans have a long-standing tradition of invoking their boundless imagination to do what humans do best: make stuff up.

Virtually every culture in the history of humankind has a creation story, a fanciful tale of how our world – all its creatures, features, the land itself as well as the sky above and the ocean beneath – came to be. Among the oldest peoples on the planet are the San of the Kalahari, who have a fascinating story about a prominent feature of the night sky. A girl undergoing her first menstruation was placed in isolation, as per custom. But there was none of the prescribed beetroot available to eat; the hunters had been gone for days and were feared lost. The hungry and annoyed girl, imbued with the powerful magic of coming into womanhood, stormed over to the campfire, scooped up its ashes and tossed them high into the sky. These glowing embers lit the way for the hunters to return home, and she and the tribe were saved. To this day these magical ashes provide a glow on moonless nights that we now call the Milky Way.

Further north, the Kuba people of Central Africa worshipped the Creator God Mbumba, who was apparently beset with stomach discomfort. On his first vomit, he created the Sun, which dried up the oceans enough for some land to appear. Unrelieved, the bilious Mbumba then proceeded to vomit up the Moon and stars, and even nine indigenous animals who then went on to spawn all the other animals in a vague parallel to Darwin's evolution.

Creation stories like these abound around the world, often fancifully incorporating local fauna. Turtles, for instance, feature prominently in Indigenous American creation stories. Such stories almost always invoke an uncreated Creator or unmoved Mover who spawned all that we see around us. So ingrained is this idea of a Creator in human consciousness to this day that its inconsistent logic usually goes unchallenged except for occasional sardonic references to 'turtles all the way down'.

From all these stories, which have come from every corner of the world, two clear messages emerge: first, humans have always looked to the heavens with a sense of awe and wonder; and second, humans have an innate need to understand their origin and place in the cosmos. These instinctive needs survive deep within us even in the modern day.

Beyond creation myths, early humans realized that the predictability of the heavens was ideal for practical uses such as timekeeping and navigation. In the ancient civilizations of both Mesoamerica and Africa, the Pleiades constellation was regarded as a marker for harvesting crops; the Aztecs even built their entire calendar around the movement of the Seven Sisters in the sky. The later Maya calendar not only tracked the Pleiades (along with the Sun and Moon), but even accounted for the heavens' 25,772-year cycle, owing to the precession of the equinoxes. The ancient Egyptians, motivated to accurately predict the flooding of the Nile, invented the first 365-day calendar, with (sensibly) twelve months of thirty days each and then a short five-day intercalary month. It is awe-inspiring to think how many centuries and millennia of carefully recorded knowledge about the night sky, passed down through countless generations of ancient astronomers from the Far East to the New World, were required in order to notice and record patterns corresponding to the precession of the equinoxes or 365 days in a calendar year. This further testifies to the central place that the skies held in ancient societies.

Throughout ancient cultures, sky knowledge was seen as being in touch with powers beyond our mundane human existence. As so often happens, with knowledge comes power. Astronomical knowledge was often co-opted by society's powerful elites in order to impress and control the plebeians. While this added to the oppression of peoples – a situation endemic in the pre-modern era – a silver lining was that it spurred those in high places to encourage and promote scientific progress. For example, in China in the third millennium BCE astrologers were tasked with predicting solar eclipses, which were regarded as important omens by the emperor. After two unfortunate fellows miscalculated the forecasts and were rudely beheaded, subsequent astrologers rapidly became quite good at this form of divination, making China the first civilization to accurately predict both solar and lunar eclipses. Such were the drivers of scientific advancement in ancient times.

Astrologers played an important role in numerous other ancient cultures whose echoes reverberate to this day. The Babylonian form of astrology and its associated constellations has survived into modern Western cultures. Unfortunately, unlike the later Maya, the Babylonians did not know about the precession of the equinoxes, so today's astrologers still use the Babylonian associations between the months of the year and the sun signs even though these are thousands of years out of date. This counterfactuality does not seem to deter the throngs of people around the world who swear by the characteristics of their astrological sign. Humans' need to connect to the heavens has been around for our entire existence, and evidently has yet to wane, even in the light-polluted modern era.

THE ROAD TO SCIENCE

The common association of the heavens with the divine naturally led to sky knowledge and its associated power being hoarded by each culture's dominant religious order. Knowledge of the sky and its workings, along with all its accoutrements such as creation stories, became absorbed into the prevailing religion in just about every culture.

Religious dogma does not necessarily play well with curiosity-driven science. More often than not, in the ancient world and even in modern times, the loser in this stand-off has been curiosity-driven science. The creation story of the cosmos was written down in ponderous tomes in the holy book of each faith as an unquestioned and incontrovertible truth. In Western religions, God or the gods created the Universe and everything in it essentially as we see it today. Eastern religions tend to favour a cyclical Universe which periodically re-emerges from the previous one. In all cases, the Universe is regarded as a static venue, the heavens are immutable and the motions of the sky reflect God's perfect machinations for all of eternity.

The ancient Greeks expressed this perfection via models consisting solely of perfect circles. This likely reflected their innate belief that we *anthropos* are the favoured creatures of the gods, and since our lives consist of pointlessly going round and round without getting anywhere, the gods must surely love their circles. Virtually all religions agree, in their pious humility, that humans are the most important creatures in all of Creation and hence all the heavens must surely revolve around our home, Earth.

This was unquestioningly accepted, until a dreamy lad named Apollonius of Perga spent hours gazing up at the pre-dawn sky when he should have been milking the sheep. He realized that circles alone couldn't fully explain the motion of a few odd points of light that wandered around the sky, the *planetes*. Apollonius dreamed up a variation on perfect circles, called epicycles. Epicycles were little circles of motion superimposed upon the big circles, which could explain why the big circles weren't perfect, while not breaking the circles motif the Greeks had going on.

The famous astronomer Hipparchus of Nicaea heard about Apollonius, and after visiting him to contrapuntally discuss the nuanced intricacies of epicycles over a cup of warm sheep's milk, stole his idea. Karma would come around though, as years later the Roman astrologer Ptolemy would effectively do the same to Hipparchus. Today, we call the idea of geocentric epicyclic motion the Ptolemaic model.

The Ptolemaic system emerged from the ancient world as the accepted cosmological model, and it was subsequently adopted by the Christian Church as the true and definitive Word of God. In the Western world, the Ptolemaic model reigned as the ordained gospel throughout Europe's Dark Ages, unchallenged except under severe penalty, as that braggart Galileo would discover.

The Renaissance brought wider literacy and freer thinking, always a dangerous combination. As the Church waned in power, there emerged a more systematic approach towards uncovering the secrets of the heavens involving observation, logic and mathematics. At this, the general public quickly lost interest.

Nevertheless, Nicolaus Copernicus, a fifteenth-century Polish mathematician and ecclesiast, dabbled in measurements of planetary motion, and encapsulated his findings in a model in which all the planets, including (heretically) Earth, revolved around the Sun. He realized that he should probably keep this revelation to himself if he preferred his head to remain atop his shoulders. His treatise, *On the Revolutions of the Heavenly Spheres*, was disseminated posthumously by alleged friends of his, who for their part disavowed any knowledge of how hundreds of copies suddenly appeared in their possession from some fellow they had never heard of before.

Old beliefs die hard, however. The wealthy king of Denmark caught wind of Copernicus' book, and like powerful people to this very day, he figured he could discredit any scientific finding if he threw enough money at it. He hired the astronomer Tycho Brahe and bequeathed him approximately 1 per cent of Denmark's gross domestic product, equivalent to about £22 billion today, to build an observatory on a windswept island hosting a rare feature of Danish terrain – a hill – and use it to prove the Church-ordained Ptolemaic model.

Sadly, Tycho's mission failed. Although over many years he succeeded in cataloguing the motions of stars and planets to unprecedented precision, he found distressingly persistent deviations from God's Plan. Tycho instead proposed the bizarre Tychonic model, in which the planets and comets all revolved around the Sun, but the Sun and Moon somehow orbited around Earth, just as God ordained.

While the king of Denmark was busy trying to make heads or tails of these cutting-edge scientific findings, Tycho surmised that an extended holiday abroad might be wise. In Poland, he took on a young German apprentice named Johannes Kepler to continue analysing his voluminous data. After Tycho's curious and untimely death in 1601, Kepler, freed from the constraints of the Danish funding agency, noticed that Tycho's observations all fitted together tidily if one assumed that all the planets including Earth orbited the Sun in ellipses rather than circles.

A few decades later, Sir Isaac Newton would explain Kepler's empirical laws of planetary motion, along with essentially all other catalogued astronomical observations at the time, as a natural consequence of his spiffy new theory of gravity. Today, other than a few minor corrections due to some upstart named Albert Einstein, Newton's laws of gravity and motion provide the basis by which numerical cosmologists simulate the cosmos.

ECHOES OF THE COPERNICAN REVOLUTION

The Copernican Revolution, as it would come to be called, was about much more than the geocentric versus heliocentric model – it was an epistemological change in the way of determining knowledge. Today, cosmology, and science in general, implicitly follow two tenets which trace their origins to Copernicus:

(i) An understanding of the workings of the world around us is best gained by building models that most simply and elegantly explain all the observations; and
(ii) Earth does not occupy a special place in the cosmos.

The first tenet usurps the idea that religious edicts represent the essence of truth, and instead casts science as the arbiter of the 'best current explanations' for how the world works. It posits that scientific knowledge is inherently mutable and challengeable, and indeed that this is an innate strength which elevates the veracity of scientific knowledge over the rigid beliefs of religion. In the Copernican view, science does not concern itself with absolute truths, but rather with developing models that best describe what we see around us, to be tested and refined towards greater pre-dictive accuracy, much like the Chinese eclipse foretellers of old. As you will see in this book, developing models to more accurately predict observations of the Universe will be a driving motif as we work towards simulating the cosmos.

The second tenet is the origin of the Copernican principle, which generalizes Copernicus' original statement to the broader idea that the Sun is not a special star, our Milky Way is not a special galaxy and perhaps even that our Universe is not a special universe. Later, Darwin would pile on by claiming that humans are not a special species. With the Copernican principle, the fall from grace of the *anthropos* is complete.

So it was that from the Renaissance through the Industrial Revolution, cosmology insidiously crept its way from the bastion of religion towards the realm of science. Yet for all its upheavals, the Copernican Revolution did nothing to dispel the notion of an eternal and unchanging Universe of divine creation. This seemed as sure as the Sun rising in the East on its daily sojourn round the Earth.

Overturning this final piece of cosmological dogma would have to await the early twentieth century. It was then that, for the first time in human history, cosmology blossomed into a fully fledged scientific endeavour. It began with a simple yet shocking observation that reverberated throughout science and society with such force that one might even call it a big bang.

It is here that our story begins.

1

THE COSMOLOGICAL FRAMEWORK

The advancement of cosmology over the past century is among the crowning achievements of modern science. While other sciences may have more of an impact on our day-to-day lives, the development of the modern cosmological framework represents a fundamental transformation in the way humans view the Universe and everything in it. It tackles questions that, not a century ago, were widely thought to be beyond the reach of human comprehension. The ongoing revolution in cosmology has today provided a well-accepted framework to help us understand why the Universe looks the way it does. This cosmological framework sets the stage for how we simulate universes on a computer.

FROM CHAOS TO CONCORDANCE

When I first entered into cosmology as a postgraduate student some thirty years ago, the field was like the intellectual Wild West. It was full of a huge number of competing theories claiming to capture the true nature of our Universe, each seemingly invented for no reason other than to sound crazier than the last. Hypotheses such as topological defects, mirror universes, cosmic strings and other bizarre notions that sounded more like Isaac Asimov than Isaac Newton were wafting through the cosmological community. Hard data on the distant Universe was sparse; telescopes simply weren't powerful enough to probe sufficiently far out into the cosmos to rein in wayward theories. Big personalities were the order of the day, and with every new bleeding-edge observation of some distant fuzzy blob, there was a cadre of cosmologists bombastically declaring that this unequivocally demonstrated their favoured model was correct. Even the most basic parameters of our Universe were a mystery: how much mass does it contain? What is it made of? How long ago was its origin? What is its fate?

Fast-forward to today, and the situation is almost unrecognizably different. Modern cosmology is now the ambit of the staid statistician, who assembles an overwhelming number of observations of the cosmos into precise constraints on the exact constituents of our Universe and how they have evolved since the Big Bang. We now know the total mass-energy density of our Universe to within a couple of per cent. The well-accepted concordance cosmological model posits that about 70 per cent of the Universe is in some mysterious form called dark energy, which acts like a pressure, causing each piece of empty vacuum to exert an unmeasurably small outwards force that, over the vast volumes of the cosmos, adds up to drive an acceleration of the cosmic expansion rate. An additional 25 per cent of the Universe is in another mysterious form dubbed dark matter, which isn't composed of the familiar protons, electrons and neutrons that, as we are all taught in school, represent the building blocks of matter. This leaves only about 5 per cent of its mass in the form of familiar periodic table elements such as hydrogen, oxygen and carbon. And of that 5 per cent, roughly three-quarters is hydrogen, one-quarter helium and a smattering of perhaps 1 per cent is all the other elements in the periodic table such as life-supporting carbon and oxygen. We, and everything we can see with our eyes, are truly the 1 per cent!

The transformation of cosmology in recent decades from a science of rampant speculation into one of statistical precision is among the most fascinating scientific journeys of the early twenty-first century. In many ways it mirrors the transformation that physics underwent in the early twentieth century, when the development of quantum mechanics and relativity reshaped our understanding of the natural world. One difference is that progress in cosmology has been driven in large part by technological advances, both in terms of telescopes that can now survey large swaths of the night sky to unprecedented depths, and computing technology that enables the processing, analysis and interpretation of such voluminous data.

Today, computers are an integral part of astrophysics research – indeed, most working astronomers and astrophysicists spend the vast majority of their day working on computers. With the accelerated advancement of computing technology over the last fifty years, an entirely new technique has emerged within the astrophysical community, complementary to observational studies with telescopes and traditional theoretical studies using pencil and paper: numerical simulations. Simulations of the cosmos have become an indispensable tool for capturing the complexity of how the Universe and everything within it evolves over time. To understand how such numerical models of the universe work, we must first understand the framework that they are trying to model. So let us explore the emergence and development of the concordance cosmological model.

THAT ONE TIME WHEN SPACE WENT BANG

Modern cosmology began in earnest in the 1920s, when Edwin Hubble, a professor at Caltech, discovered that all sufficiently distant galaxies in every direction are receding away from us. He did so by using Doppler shift measurements, that is, the change in the colour of light emitted by a luminous object speeding away (called 'redshifting'), calculated by his colleague Milton Humason and which gave the recession velocities, combined with a technique to estimate distances developed by Henrietta Leavitt using pulsating stars called Cepheid variables. Hubble's work was made possible by having access to (at the time) the largest telescope in the world, the 100-inch telescope atop Mount Wilson above Pasadena. By graphing the recession velocity versus the distance, Hubble noticed that the further out one looks, the faster the galaxies are receding from us.

Hubble's seemingly innocuous discovery had shocking implications. The most basic of these was that the Universe was not unchanging, as had been taken for granted in both the scientific and popular communities. The Belgian Jesuit priest and theoretical physicist Georges Lemaître reasoned that, if one extrapolates backwards in time, Hubble's data implied that at some long ago time, the Universe was compressed into a single point. This would be the beginning of our Universe as we know it: the Big Bang.

The notion of the Big Bang is unlike explosions we are familiar with here on Earth. We usually think of explosions as propelling material outwards from a point, like a blast wave from a bomb. This is expansion *in* space. The expansion of the Universe, and the Big Bang, is not an expansion in space – it is an expansion *of* space.

Illustration 1 shows the difference between these two types of expansion. Our galaxy, the Milky Way, is depicted as the blue

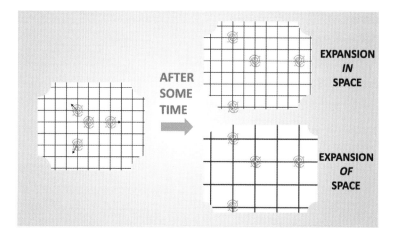

1 Expansion *in* space versus expansion *of* space. Over time, red galaxies move away from the blue galaxy. An expansion in space is depicted in the top right diagram, where each red galaxy is now double the number of grids away. An expansion of space is depicted in the lower right, where the grid itself has expanded. Explosions from bombs are the former; the expansion of the Universe is the latter.

galaxy, with all red galaxies moving outwards away from us; this is what Hubble observed. After some time, the red galaxies are all seen to be twice as far away. In contrast, for an expansion *in* space, the galaxies have moved with respect to the underlying grid (marking out space). In this scenario, there is only one galaxy that has not moved, and that's us! This puts us in a very special place in the cosmos, effectively at the centre of the Big Bang, which is distinctly anti-Copernican.

To avoid angering the ghost of Copernicus, Lemaître interpreted the outwards motion of galaxies as an expansion *of* space. In an expansion *of* space, the underlying grid itself grows. In this scenario, depicted as the bottom branch of the illustration, you will notice that each galaxy still views itself as being on the same grid point as before; for instance, the red galaxy to the right of our blue one is two grid points away from us in the first part of this diagram, and still two grid points away in the second. However, the grid points themselves are now further apart.

This might seem like a strange notion. How can space itself expand? It turns out, space can not only expand, but deform, shrink and warp. Just a few years earlier than Hubble's discovery, in 1915, Albert Einstein came up with the idea of general relativity, which argued that space is deformed by the presence of mass; the effect of such deformations is what we call gravity. He further argued that the way that different observers view the passage of time is also impacted by gravity, so it isn't only a deformation of space but is more properly viewed as a deformation in *space-time*.

Space-time is a much-abused concept, with science-fiction writers often invoking the warping of it to flit spaceships about at will. But space-time is not a physical entity, any more than the longitude and latitude lines on your world map. Instead, it is a mathematical construct, by which we measure the distance between locations in space at a given time, known as events.

In cosmology, space-time expands. The mathematics of this was worked out by Lemaître, along with his contemporaries Alexander Friedmann, Howard P. Robertson and Arthur Geoffrey Walker; today we call this framework the FLRW metric (metric being the mathematical representation of distance measures). They noticed that in this situation, like in our cartoon diagram, observers in each pocket of the Universe will view their own galaxy as stationary in space. And they will also see other galaxies receding. That is to say, the distance between objects at rest in their own reference frame increases with time. The expansion of space-time is like a diverging current, carrying boats away from each other even though each boat sees itself as stationary with respect to its surrounding water.

From illustration 1, it is clear that the further away the galaxy, the faster its recession rate will be. This is exactly what Hubble observed. In cosmology today, this is expressed as the Hubble–Lemaître Law:

$$V = H_0 \times D$$

V is the recession velocity of a distant object, D is the distance to the object and H_0 is a quantity known as the Hubble constant that sets the proportionality between distance and recession velocity – that is, the rate at which the Universe is expanding. The 'constant' part of the Hubble constant is a misnomer: the Universe has expanded at different rates in the past compared to today, so the value of H_0 has changed over time.

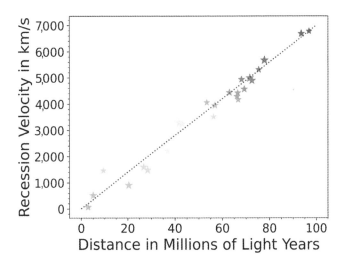

2 The Hubble–Lemaître Law is a linear relationship between the distance to a faraway galaxy and its velocity of recession. Galaxies receding at high velocity have their light waves stretched by intervening cosmic expansion, causing them to appear redshifted. Galaxies do not lie exactly on the relation, due to local gravity causing extraneous motions called peculiar velocities. An example is the galaxy Andromeda, which is approaching the Milky Way due to its peculiar velocity generated by mutual gravitational attraction.

H_0 has a measured value today of around 70 km/s per mega-parsec (illus. 2). That is, for every megaparsec further out one goes from a given location, the galaxies are typically receding 70 km/s faster. A megaparsec is a very large distance – about 3 million light years. One has to go a long way out in the Universe before the Hubble–Lemaître Law comes into play.

ON EXPANDING UNIVERSES AND PARKING SPACES

One of my professors at Santa Cruz had a T-shirt emblazoned with a whimsical yet profound question: 'If the Universe is expanding, why can't I ever find a parking space?' The answer to this question highlights another crucial aspect about cosmic expansion: gravity counteracts the expansion of space.

According to general relativity, mass bends space-time, draw-ing objects towards each other. A useful two-dimensional visu-alization of this intrinsically three-dimensional space is to think of a rubber sheet with an object on it, creating a depression that will draw other objects in the sheet towards it. In the two-dimensional sheet, the bending occurs in an unseen third dimension; in our three-dimensional space, the bending occurs in an unseen fourth dimension. But the net effect is the same – gravity causes the paths of objects to bend towards mass.

The rubber-sheet space-time around a galaxy is depicted in illustration 3. Far away, space-time is expanding. But as one approaches the galaxy in the centre of the illustration, its mass counteracts Hubble–Lemaître expansion. A galaxy thus has an *infall region* around it, within which its gravitational influence causes space-time to contract into a static, bound region of space. Within this bound region, matter is no longer expanding or contracting, but instead is orbiting in a stable configuration.

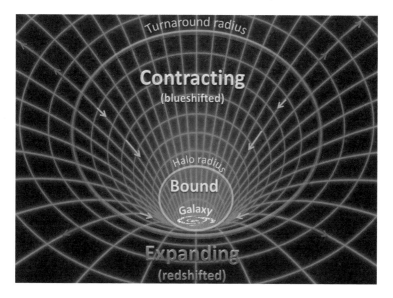

3 Local gravity counters expansion: a galaxy sits within a potential 'well' in space-time, where its gravity draws in surrounding matter. Meanwhile, on very large scales, the Universe is expanding. In between, there is a turnaround radius, outside of which space-time is expanding, and within which it is contracting. Close to the galaxy, one enters the gravitationally bound region. Everything that we see around us, the stars in the sky and nearby galaxies like Andromeda and the Magellanic Clouds, is within a bound region called the Local Group. The Local Group has broken away from cosmic expansion, and so does not expand according to the Hubble—Lemaître Law. One must go millions of light years away, well past nearby galaxies, in order to reach the expanding region.

Everything we can see with the naked eye is within a bound region of space-time. This includes Earth and everything on it, the Solar System, the stars in the Milky Way, nearby galaxies like Andromeda and the Magellanic Clouds, the atoms inside our body and yes, even parking spaces. None of these things are expanding due to Hubble–Lemaître expansion. To see the expansion of space, one needs to look on very large scales, millions of light years away. This is why it took us until the early twentieth century to be able to detect cosmic expansion, since we needed to build telescopes powerful enough to do so.

If one extrapolates the Hubble–Lemaître Law outwards, eventually the recession velocities will exceed the speed of light. Doesn't this violate Einstein's theory of relativity, which imposes the speed of light as a cosmic speed limit? No, it turns out. There is fine print in the theory that is often omitted in the sound-bite version: Einstein's universal speed limit only applies to objects moving relative to each other *in the same reference frame*. In the case of distant galaxies, they are moving in a different reference frame, one that is expanding away from us. So very distant galaxies can indeed be receding at speeds faster than light, with no violation of Einstein's edict. The catch is, we can never see these galaxies (and they can never see us) regardless of how powerful our telescopes become, because their light has not had time to reach us. Such galaxies lie beyond our cosmic horizon – the distance that light can travel to us since the Big Bang. This distance defines the radius of our 'Observable Universe'.

Our Observable Universe has a centre: us! This doesn't make us special, because everyone's Observable Universe is centred on themselves. But is there an actual centre of the entire Universe? This is not known; if it has finite size, there must indeed be a centre, but it may well be far outside our Observable Universe, and it is unclear whether this point would be special in any way. It is thus not useful to think of the Big Bang as occurring at a particular location – it happened everywhere in our entire Universe, at once. The Big Bang is not a point in space, but rather a moment in time.

AN ORIGIN STORY, TWENTY-FIRST-CENTURY STYLE

The Big Bang model postulates that the entire Observable Universe was once compressed into a very hot, very dense state. From this, purportedly, everything that we see around us emerged over time. But how did this unfold? In the century following Hubble's discovery, physicists have pieced together an intricate but convincing story, backed by both observations of the very large (using data from telescopes) and the very small (using data from particle accelerators), of our cosmic origins. Here is how the current story goes.

Lemaître argued that if one extrapolates the expansion backwards, one reaches a time when the entire Universe is compressed into a single point of zero size. When was that? The Hubble constant gives us a good estimate. Its measured value is around 70 km/s/Mpc, which is equivalent to a fractional expansion rate of 0.0000000000715 per year. By inverting this, one obtains the number of years the Universe would need to expand from a *singularity* (that is, everything condensed to zero size) to reach our current size: 14 billion years. This simple calculation shows that the Universe must be around 14 billion years old.

What was the Universe like at the very beginning? For one thing, it's worth noting that the singularity probably never happened. In physics, singularities occur when we try to extrapolate known laws of physics into unknown regimes – the result doesn't make sense. It turns out that the laws of physics as we know them fail prior to a time that is 10^{-43} seconds after this purported singularity. This is known as the Planck time. To understand why everything goes haywire at the Planck time requires knowing a bit about the four fundamental forces of nature.

All known forces in nature can be categorized into four fundamental forces: the everyday forces of gravity and electromagnetism, along with the weak and strong nuclear forces that hold atoms together and only operate on subatomic scales. Prior to the Planck time, the Universe was so dense that all four forces were unified into a single 'superforce'. While physicists have figured out how to unify the two nuclear forces and electromagnetism into what is called the Grand Unified Theory (GUT) force, folding in the force of gravity has proven to be extraordinarily elusive. Einstein worked on this for the last forty years of his life, to no avail. Arguably the greatest living physicist today, Ed Witten, has worked on this for much of his life, basing his research on the notion of string theory and his eleven-dimensional version of it, M-theory. Unifying gravity and the GUT force is perhaps the single greatest challenge in particle physics today. But as of now, no fully successful framework has emerged to mathematically describe this situation. To understand the Universe prior to the Planck time requires knowing how to unify all four forces of nature. Until this is solved, the Planck time will remain an inscrutable horizon of knowledge for modern physics.

It is unclear how the Universe got itself into this hot dense state at 10^{-43} seconds. The origin of this 'initial condition' is not part of the Big Bang model; this model makes no statement whatsoever about where the Bang came from. That doesn't mean that cosmologists are short of ideas – current notions include cyclic universes, multiverses, colliding branes or quantum foam. This is the new Wild West of cosmology, long on creative thinking but short on hard evidence. Regardless of how the Universe got itself into this hot dense state, all available evidence points to the fact that this was its state at 10^{-43} seconds. The Big Bang theory merely postulates that such a state pops into existence at that time, and then works out the consequences.

A BRIEF HISTORY OF HISTORY

The hot, dense state pops into existence. What happens next? To answer this, let's think about a Chinese lantern. At the Shangyuan

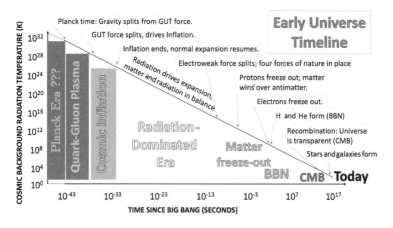

4 Cosmic timeline, shown with a logarithmic timescale to emphasize the early Universe. The temperature of the cosmic background radiation is shown along the vertical axis, starting from extremely hot to its current temperature of 2.73 K.

Festival on the fifteenth day of the first month of the Chinese year, children place a candle inside a paper enclosure, light the candle and watch the paper lantern float off into the air in a wondrously magical display. But it isn't magic, it's just physics – the candle heats the air inside the lantern, which causes its pressure to increase and air to be pushed outwards to lower its density, making it buoyant.

The early Universe is something like a Chinese lantern, only with a stupendously powerful candle. Just like the lantern, the heat generates pressure and drives an expansion which lowers the density. This further lowers the temperature, and the Universe cools down – which in turn lowers the expansion rate. So in the early Universe, cosmic expansion starts out incredibly fast and immediately begins slowing down.

The temperature of the Universe is a useful barometer for the time shortly after the Big Bang (see illus. 4). Changing temperatures cause phase transitions – like ice turning into water turning into steam. Phase transitions release energy, such as when ice cracks. This is because there is energy locked into the symmetry of the ice, and when it becomes free-flowing water, this symmetry is broken, releasing some of the energy.

At a time of 10^{-35} seconds, the first cosmic phase transition happened. At this time, the GUT force split into the strong nuclear force and the combined electro-weak force, leaving three forces of nature (gravity, strong and electro-weak). This is a type of symmetry breaking: prior to this phase transition, the electro-weak and strong forces were interchangeable and therefore symmetric, but afterwards, they were not.

The energy released from this phase transition drove the first important phenomenon we encounter after the Big Bang: cosmic inflation. The sudden energy input ignited an unimaginably rapid expansion of the Universe, not dissimilar to a second Big Bang. During a very brief phase that lasted perhaps 10^{-32} seconds, the Universe expanded by at least a factor of 10^{20}. That's equivalent to taking an atomic nucleus, and in the blink of a mosquito's eye, expanding it to a size 100 times bigger than Earth! As we will see later, it was during cosmic inflation that the seeds of everything we see around us – all the galaxies, stars and even planets like Earth – were sown.

During this very early epoch, still well within the first second of existence, the Universe was a plasma of energetic particles along with packets of light energy called photons. At around one-millionth of a second (10^{-6} s) after the Big Bang, the Universe became sufficiently cool that a particularly important particle, namely protons, started 'freezing out'.

To understand freeze-out, recall Einstein's famous formula, $E = mc^2$. This formula relates the rest mass of an object (say, a proton) to its rest energy. Imagine two energetic photons smash together. Occasionally, these photons will spontaneously create a proton–antiproton pair, since they will have enough energy to do so. The proton and antiproton will eventually find their counterpart and annihilate back into photons, creating an equilibrium, that is, a balance between the number of photons and protons/antiprotons.

But once the Universe's energy drops below the rest energy of protons, the typical photon no longer has enough energy to create proton–antiproton pairs – yet those remaining pairs can still annihilate. So all the protons and antiprotons annihilate, but the photons are no longer energetic enough to create new protons and antiprotons. This breaks the equilibrium, leaving many photons and very few protons or antiprotons.

One might rightly ask, why don't they *all* annihilate? Why is our Universe filled with protons, such as in the atoms of our body and those that make up Earth, but almost no antiprotons? Why did matter win out over antimatter? It turns out that there is a tiny imbalance in the weak force known as charge-parity (CP) violation – once in every billion or so photon collisions, a proton is produced *without* an associated antiproton. As such, prior to freeze-out, there aren't an equal number of protons and antiprotons; there are one-part-in-a-billion more protons than antiprotons. When freeze-out occurs, all the antiprotons annihilate, but this leaves those lucky one-in-a-billion protons left over. It is due to CP violation that we have a Universe containing matter, not antimatter.

Moving on to around fifteen seconds after the Big Bang, electrons similarly freeze out, with CP violation again leaving one electron for every billion electron–positron pairs. Incidentally, there's good reason to believe that the dark matter particle, whatever that is, froze out before protons. As such, after about fifteen seconds, the Universe was full of protons, electrons, dark matter and a smattering of more exotic particles. For every such particle of matter, there are about a billion photons flying around, and this is owing to CP violation. These photons fill space with energy, like a remnant echo of the Big Bang. This remnant echo turns out to be one of the most important ways in which cosmologists can measure and understand our Universe. We call this remnant echo the cosmic microwave background (CMB).

THE BIG BANG MODEL WINS THE DAY

A single proton is identical to a hydrogen atomic nucleus. You might recall that the nuclei of anything heavier than hydrogen in the periodic table contain neutrons. This requires that the Universe creates neutrons and then sticks them together with protons to form a nucleus. In a fortunate coincidence, during the period between one minute and three minutes of cosmic time, the Universe was precisely hot and dense enough to assemble two protons and two neutrons into helium atoms. During those two eventful minutes, the Universe put about 25 per cent of the mass into helium nuclei, while the other 75 per cent were still free-floating protons, that is, hydrogen nuclei. This process is known as Big Bang nucleosynthesis (BBN).

BBN was published as a natural consequence of the Big Bang model by Ralph Alpher and collaborators in the 1940s, and was considered a huge triumph for the Big Bang model. Before BBN, it was a complete mystery to astronomers as to why stars like our Sun all seemed to be made up of about three-quarters hydrogen and one-quarter helium. Thanks to BBN, it was recognized that this composition arises directly from what happened during the first three minutes after the Big Bang.

The Universe remained plasma for many years after, with free electrons and protons mixed with CMB photons. Protons and electrons could not yet bind into neutral hydrogen atoms, because there were still too many energetic photons flying around. Any nascent pairing of a proton and electron was soon smashed by an overzealous photon, tearing those poor soulmates apart like sub-atomic Romeos and Juliets. This plasma state continued for about 380,000 years. At that point, the temperature of the Universe had cooled to about 5,000 Kelvin, similar to the surface of the Sun. The energy of the cosmic photons is, at this temperature, similar to the energy it takes to unbind an electron from a proton in a hydrogen atom. So once the temperature drops below this, if an electron and proton form into a neutral hydrogen atom, the typical cosmic photon no longer has enough energy to smash it apart. As a result, stable hydrogen atoms in the cosmos can finally form. This event is called 'recombination'.

Recombination is hugely important, because charged plasma is opaque to light but neutral gas (such as air) is mostly transparent. So at 380,000 years the Universe went from opaque to transparent, and the photons started streaming freely throughout the cosmos, a journey that continues to this day.

Owing to cosmic expansion, the CMB has now cooled down to 2.73 K. The CMB was first spotted serendipitously in 1964 by two radio astronomers, Arno Penzias and Robert Wilson, who were trying to develop better satellite communication receivers for Bell Labs; they won the Nobel Prize in Physics for this discovery in 1978. Following Edwin Hubble's discovery of the expanding Universe in the 1920s and Alpher's success with predicting the abundance of hydrogen and helium in the 1940s, the discovery of the CMB was the clincher for the astronomy community to accept the Big Bang model. Since then, evidence in its favour has continued to mount ever more quickly, so that today there are no viable competing models.

ASTRONOMERS MISPLACE 90 PER CENT OF COSMIC MASS

What is the Universe made of? This question seems like it should have an easy answer. You probably learned in a chemistry lesson at school that matter is made up of the elements in the periodic table, which themselves are made of protons, neutrons and electrons. Electrons are fundamental, but particle physicists tell us protons and neutrons are made of even smaller subatomic particles called quarks and gluons. Sprinkle in a few more strange beasties such as neutrinos and the Higgs boson, discovered in 2012 at the European Laboratory for Particle Physics (CERN), and this rounds out the so-called Standard Model of particle physics. According to the particle physicists in the mid-twentieth century, everything we have ever seen is made up of particles in the Standard Model.

Not so fast, said the cosmologists. According to them, not only is everything not made up of elements in the periodic table, but all that stuff combined only makes up about one-sixth of the mass in the Universe. The rest is in the form of some mysterious dark matter, which is in all likelihood a subatomic particle not found in the Standard Model at all.

At first, the particle physicists were a bit miffed. It's as if someone takes their car in to a garage to get it repaired, the technician runs a full diagnostic and suggests a solution to get it running again, but the customer retorts, 'Nah, I read some stuff on the Internet last night, and I'm pretty sure the problem is that my car is missing five-sixths of its engine.' Eventually, however, the particle physicists came around when they realized that searching for this mysterious dark matter might be quite a profitable venture. And there is no better way to soothe an academic's indignation than some grant funding.

Dark matter is defined as a substance that does not interact with light – no emission, no absorption, no reflection or refraction. But it has a lot of mass, which Newton and Einstein both told us creates a lot of gravity. Gravity causes things to speed up by pulling objects towards bigger objects. So by measuring how much mass we can attribute to visible matter, and comparing that to how much gravity is seen via the speeds of objects, we can indirectly look for missing mass.

Fritz Zwicky was the first to use this idea in 1932 by measuring the speeds of galaxies moving around in the Coma Cluster, and comparing that to the visible mass (in stars) of those galaxies. He found that there was a lot of missing mass needed to explain the motions, which he called *dunkle Materie*, or dark matter. Soon after, Jan Oort measured the speed of stars near the Sun, and suggested that there was even missing mass right in our own cosmic backyard. These observations were tantalizing but unconvincing, so the community remained sceptical.

Fast-forward to the 1960s, and Vera Rubin applied the same idea to stars and gas rotating in the tenuous outskirts of disc galaxies. In the furthest outskirts Rubin found five to ten times more mass than expected, which was subsequently confirmed in many more galaxies. This led to her now-famous quote, 'In a spiral galaxy, the ratio of dark-to-light matter is about ten. That's probably a good number for the ratio of our ignorance to knowledge.'

Rubin was equivocal in her interpretation of the data. She acknowledged Zwicky's dark matter idea, but also offered the possibility that perhaps Newton and Einstein were wrong, and that gravity works differently in the deepest reaches of space. To this day, a handful of scientists continue to explore such modified gravity models. But while such models can explain Rubin's data quite well, an avalanche of independent data across all scales and cosmic epochs that strongly favours Zwicky's dark matter has emerged in recent decades. Today, dark matter is by far our best current explanation for the missing-mass phenomenon, essentially universally accepted by the astrophysical community.

MACHOS VERSUS WIMPS

To astronomers, the most obvious explanation for the missing mass was that it was ordinary matter in some undetectable form, such as faint stars, giant planets or black holes. These are generically referred to as massive compact halo objects (MACHOS). Several groups cleverly devised a way to look for MACHOS using a technique called 'gravitational microlensing' (illus. 5).

The idea is that if the Milky Way is chock-full of dark MACHOS, occasionally one would pass directly in front of a bright star. By Einstein's theory of general relativity, the MACHO would bend the light paths around the MACHO towards Earth, creating a sudden brightening of the background star like a magnifying glass focusing sunlight, followed by a dimming as the MACHO passed on. By searching for such microlensing flares, it would be possible to 'see' the invisible objects, and even measure their mass. Sure enough, many MACHO events were seen – but in adding up all their mass, it was soon recognized that there were far too few MACHOS around to explain dark matter.

Meanwhile, the particle physicists had their own angle. From their point of view, dark matter was a subatomic particle with two characteristics: (i) it was massive; and (ii) it did not interact electromagnetically. Astroparticle physicists proffered that dark matter was made up of WIMPs – weakly interacting massive particles. The mid-twentieth century was heady days in the particle physics community, with increasingly powerful accelerators discovering new subatomic particles almost daily. Surely one of these days, they thought, they would run across a WIMP with exactly these properties, solve dark matter and triumphantly lord it over those silly astronomers. But to this day, they too have found nothing conclusive, even when co-opting entire underground mines to construct elaborate dark matter detectors. With WIMPs looking shaky, particle physicists are now exploring a much wider range of dark matter candidates, including axions, sterile neutrinos and light supersymmetric particles.

In the early 1970s, cosmologists jumped on the dark matter bandwagon. If there was ten times more mass in dark matter than

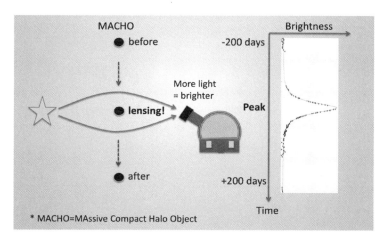

5 Schematic of microlensing searches for massive compact halo objects (MACHOs). As an otherwise-invisible MACHO moves in front of a background star, the light from the star is gravitationally focused onto Earth like a magnifying lens, causing the star to brighten temporarily in a very characteristic pattern. The brightness data shown in the right graph is an example from the Optical Gravitational Lensing Experiment (OGLE, star OGLE-2005-BLG-006) taken over more than four hundred days, showing the star's brightness peaking during a microlensing event. From the shape of the light curve, the mass of the foreground MACHO can be estimated.

ordinary matter, then surely it would affect the evolution of the Universe and the objects within it. But cosmologists can't help but stoke infighting. Two camps fought it out: the theoretical cosmologists preferred a value of about 20 to 50 for the ratio of dark to ordinary matter. This would elegantly make the Universe have just enough mass to slow down the expansion rate to zero, but not enough to re-collapse. The observational cosmologists meanwhile didn't give a whit about elegance, instead preferring cold hard data that suggested a dark-to-ordinary matter ratio closer to 5. The battle was on.

The definitive resolution to all these debates came from a surprising source: the echo radiation of the Big Bang, the CMB. It turns out that the CMB precisely tells us not only the nature of dark matter, but the ratio of dark to baryonic matter in the Universe. Baryonic matter is what astronomers call all the matter that interacts with light – that is, ordinary matter made up of protons, neutrons and electrons. This ratio will be crucial for simulating the cosmos, so let's take a deeper dive into how the CMB tells us this.

THE SYMPHONY OF THE CMB

The CMB streams to us from 380,000 years after the Big Bang, when neutral atoms were first able to form and the Universe became transparent. After the inflationary epoch, the Universe was seeded with extremely tiny fluctuations in mass (we'll see why later). The regions with ever-so-slightly more mass attracted mass around them, and slowly grew into larger fluctuations; this process is known as gravitational instability. But by 380,000 years, the matter fluctuations were still tiny: less than one part in 100,000, smoother than the stillest pond on Earth.

One can think of these mass fluctuations as peaks and valleys. A CMB photon starting out in a region with slightly more mass effectively has to climb out of a tiny valley against gravity, losing a bit of energy relative to one that came from a region with less mass; this is known as gravitational redshifting, and is another consequence of general relativity. This means that CMB photons from the valley will appear slightly cooler to us, relative to the typical CMB photon. Conversely, a photon that started out in a less dense region will appear hotter than average. This is how the temperature fluctuations in the CMB reflect mass fluctuations present in the Universe at that time.

Illustration 6 shows an all-sky measurement of these primordial temperature fluctuations measured by the Planck satellite, a CMB probe launched by the European Space Agency in 2009, that were obtained after carefully subtracting away numerous foregrounds such as dust emission in the Milky Way and the motion of the Earth relative to the CMB. Blue regions are cooler, containing a-few-parts-in-a-million more mass than average, while red regions are hotter and have less mass. In essence, this is a baby picture of our Universe at the ripe young age of 380,000 years.

The CMB temperature fluctuations can be quantified via a harmonic power spectrum. A harmonic power spectrum means breaking down the CMB sky into different frequencies, akin to breaking down the sound from a piano concerto into a combination of different notes of varying strengths. Amazingly, much like a piano, the early Universe prefers to ring at certain frequencies. This ringing is known as baryon acoustic oscillations (BAO). It turns out that the BAO is an absolute gold mine of information about the constituents of our Universe. Let's see why.

Before 380,000 years, the Universe was still an opaque plasma. This means the photons and baryons were coupled

6   Planck CMB fluctuation map. This is an all-sky map of the cosmic microwave background fluctuations, which reflect the matter fluctuations imprinted during cosmic inflation. This shows the seeds of today's stars and galaxies.

– where the photons went, the baryons were dragged, and vice versa. Meanwhile, the dark matter doesn't interact with photons, so it did its own thing. As gravitational instability progressed dark matter collected together and the density fluctuation grew. The baryons were also attracted by gravity, but as the baryons collected together, they brought the photons with them. When photons were squeezed, the pressure created pushed back against gravity, carrying the baryons back outwards. The baryons were thus caught pulsating between the dark matter's gravity pulling inwards and the photon's pressure pushing outwards. The net result was an oscillatory 'ringing' of the baryons within these nascent matter clumps – these are what cosmologists call the baryon acoustic oscillations.

Illustration 7 shows the harmonic power spectrum of the Planck CMB map in illustration 6. Those huge peaks come from the BAO. These are the characteristic frequencies at which the 380,000-year-old Universe likes to ring. The first harmonic has a wavelength of 380,000 light years, which corresponds to an angular scale of about one degree on the sky. This represents the largest causal region that can collapse by an age of 380,000 years, with baryons and dark matter falling inwards together for the first time. The second harmonic represents the baryons falling in and then getting pushed back outwards versus the dark matter; because this requires both falling in and moving out, the largest causal region for this mode is 190,000 light years. And so on with higher harmonics. The BAO is the original music of the spheres.

How does the BAO allow astronomers to measure dark matter? Here's the key idea: the first peak represents baryons and dark matter falling in together, while the second peak represents dark matter falling while the baryons are being pushed out. So the difference in the strength of these first two peaks measures the amount of stuff that is moving oppositely – the baryons – relative

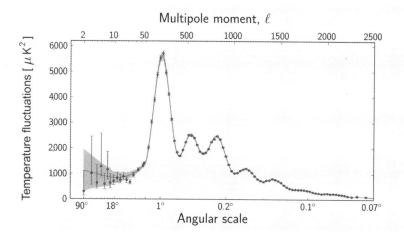

7 CMB harmonic power spectrum, showing the spectrum of the temperature fluctuation map depicted in illus. 6 decomposed as a function of angular scale (which is related to frequency). The baryon acoustic oscillations are very prominent, with the first peak showing up at about 1°, and subsequent harmonics at smaller scales. Higher harmonics have lower power, and also show an odd–even pattern where the odd peaks are strong and the even peaks are weak. The odd–even contrast enables our most precise measurement of the cosmic dark-to-baryonic matter ratio.

to the dark matter. In other words, the ratio of the BAO's first two harmonic peaks (and odd versus even peaks in general) tells you the ratio of dark to baryonic matter.

In the early 2000s, the Wilkinson Microwave Anisotropy Probe (WMAP) measured the first few BAO peaks. This very accurately estimated the mass in the Universe to be approximately five-sixths dark matter and one-sixth baryonic matter, giving a ratio of five to one. This settled one of the debates: the observational cosmologists were correct. It was also comforting that observers had measured this ratio from looking at the present-day Universe, while WMAP had measured it in the early Universe, yet the value was the same. This is an example of one of the many concordances in the concordance cosmological model: there is concordance between measurements of the dark-to-baryonic matter ratio of five from the CMB in the early Universe and the motions of objects around us today.

The BAO tells us even more than this. It demonstrates conclusively that dark matter is *not* made up of any type of particle that interacts with photons – it cannot consist of MACHOs or ordinary gas in some unseen form. This is because the BAO indicates that dark matter does not experience photon drag, which explicitly shows that it does not interact with light. Hence dark matter cannot be made of protons, neutrons and electrons like the elements in the periodic table, all of which interact with photons. This settles the other debate: the particle physicists won out over the astronomers.

The evidence for dark matter from numerous independent measurements of the Universe continues to grow, and it is now a secure part of the concordance cosmological model. At this point, Zwicky and Rubin's 'missing mass', despite leading to dark matter's discovery, is not even close to the strongest piece of evidence for dark matter. While terrestrial dark matter searches have not identified the dark matter particle(s), this isn't considered a strong argument against its existence; many particles in physics were conclusively inferred long before their actual discovery, such as the neutrino and the Higgs boson. It merely shows that humans lack the technology to directly detect dark matter at this time – or perhaps ever. After all, the Universe is under no obligation to make all her secrets accessible to us puny humans.

THE FLAT UNIVERSE SOCIETY

For cosmologists, the BAO is the gift that keeps on giving. It turns out, the frequency (or angular scale) of the first peak in the CMB harmonic power spectrum also gives another important clue about the nature of the cosmos: our Universe is flat.

Now before you think that cosmologists are some spaced-out chapter of the Flat Earth Society, let me point out that 'flat' perhaps doesn't mean what you think it does. It doesn't mean the Universe is shaped like a pancake, or the Universe has an edge that you can fall off or that the BAO is a NASA hoax. Here, 'flat' is a term borrowed from differential geometry meaning that the Universe obeys the rules of Euclidean geometry.

Euclidean geometry is defined by the rules that you know and hate from school, such as 'two parallel lines never meet' and 'the angles of a triangle add up to 180 degrees.' But there are other possible geometries, such as closed or open. As an example of a closed geometry, consider the Earth's surface. On a globe, two parallel lines at the equator (longitude lines) meet at the poles, while a triangle's three angles add up to more than 180°. Such a geometry does not obey Euclidean rules. An open geometry is the opposite: parallel lines diverge and the angles add to less than 180°.

Going back to general relativity, remember that mass bends space. Through $E = mc^2$, energy can equivalently bend space. So it shouldn't be surprising that the geometry of the Universe is connected to its total mass-energy density. A high density gives a closed Universe, while a low one gives an open Universe. In cosmology, this is quantified by the parameter Ω: a value of $\Omega > 1$ corresponds to a closed Universe, $\Omega < 1$ an open Universe and $\Omega = 1$ a flat Universe, the last of which occurs at a very special critical density. Today, the value of this critical density is about 10^{-29} grams per cubic centimetre, roughly five hydrogen atoms every cubic metre. This is the typical density in deep intergalactic space, which is far emptier than any 'vacuum' that humans have produced on Earth.

The fate of the Universe is also connected to its geometry. This can be envisioned by two objects moving on two tracks: in a closed Universe, the tracks might be diverging now as we move away from the South Pole towards the North, but eventually the two tracks will converge; thus a closed geometry corresponds to a Universe re-collapsing in a so-called Big Crunch. For an open geometry, the Universe doesn't have enough gravitational mass-energy to re-collapse, so the Universe keeps expanding, and the objects' paths diverge forever. A flat geometry is the Goldilocks case between open and closed, corresponding to a mass-energy density that is just enough to slow the expansion rate to zero, but not enough to re-collapse; two parallel tracks remain parallel forever, corresponding to a Euclidean geometry.

The location of the BAO's first peak tells us that the Universe is flat, with $\Omega = 1$ (to within a couple of per cent), meaning that the Universe's total mass-energy density is precisely the critical density. But the connection of geometry with the fate of the

Universe turns out to be complicated by another unexpected twist: dark energy.

EINSTEIN WAS SO WRONG HE WAS RIGHT

Various lines of evidence such as Big Bang nucleosynthesis tell us that the Universe contains about 5 per cent of the critical density in baryonic matter. The BAO and other measures tell us that the dark-to-baryon mass ratio is about five, meaning that there is about 25 per cent dark matter. That makes 30 per cent of the critical density. If the Universe is flat as the CMB says, then what makes up the other 70 per cent?

The answer turns out to be dark energy. The catchy name is a play on dark matter, and its nature is even more mysterious. But as with so much other nomenclature in astronomy, its name is something of a misnomer – it is not energy in the same sense as light energy (photons). Instead, it is best described as a vacuum pressure, which acts like anti-gravity.

The idea of dark energy traces back to (yet again) Einstein. In 1915, when Einstein wrote down his theory of general relativity, he figured a good first test would be whether it predicted a static Universe. He quickly recognized that, according to general relativity, the Universe's mass should begin pulling everything inwards and collapsing. With Hubble's discovery of a non-static Universe yet a decade away, the cosmos was unquestionably thought to be static and immutable. Einstein was flummoxed, faced with embarrassment that a basic prediction of his elegant brainchild violated a fundamental tenet of the Universe.

But Einstein was nothing if not clever. He realized that he could add a single constant, which he called a cosmological constant, denoted by the Greek letter lambda (Λ), to his field equation.

There was no obvious reason why it should appear in the equation, but it was allowed, so Einstein threw it in to see what would happen.

Remarkably, Λ behaved like anti-gravity. By choosing precisely the right value, Einstein could exactly balance the inwards force of gravity from cosmic mass. Sure, it was a bit like balancing a bowling ball on the head of a pin, but it would keep the Universe static and immutable, as God ordained, and Einstein's precious theory would be saved.

A few years later, Hubble discovered cosmic expansion. Einstein facepalmed. Had he trusted his own mathematics instead of his preconceived beliefs, general relativity could have predicted that the Universe was non-static. Because of this faux pas, he would later refer, as his colleague George Gamow noted in *Scientific American* in 1956, to the introduction of the cosmological constant as his 'greatest blunder'.

Skip ahead to the 1980s. With the Universe now known to be non-static, the cosmological constant became unnecessary, relegated to textbooks as an exercise for students. But all that changed in the 1990s. For nearly a decade, two independent teams embarked on a project to use a specific type of exploding star called Type Ia supernovae (illus. 8) in order to measure the cosmic expansion rate.

A Type Ia supernovae has a predictable intrinsic brightness, which, when combined with the measurement of its apparent brightness in the sky, allows an estimate of its distance via the fact that brightness drops off as distance squared. Since supernovae are extremely bright, cosmologists could now do the same experiment that Hubble did for nearby galaxies, except out to distances halfway across the Universe. The goal was to calculate how the Hubble constant H, which measures the cosmic expansion rate, had changed over time. The natural expectation was that cosmic expansion was slowing down, as it had been since the time of the Big Bang.

What they found revolutionized modern cosmology, and ended up garnering a joint Nobel Prize in 2011: the Type Ia supernovae were dimmer than expected, even if the Universe was completely devoid of any mass at all. This implied that the Universe's expansion rate had been accelerating, not slowing down.

Why does dimness imply acceleration? An analogy might help. Imagine your friend gets in their car and drives off into the evening. For a minute you watch as their lights get dimmer. Based on that, you can predict how much dimmer those lights would be after an hour. When you look an hour later, you find that they are even dimmer than expected. This means your friend has gone further than you thought, which implies that at some point, they accelerated their car. This is the exact same reasoning used to argue why unexpectedly dim Type Ia supernovae imply cosmic acceleration, as depicted in illustration 9.

An accelerating expansion was jarring to cosmologists. It was almost as shocking as if you tossed a ball in the air, and instead of falling back down, it accelerated on up into the wild blue yonder. If this happened, you would immediately think that the ball had some sort of propulsion mechanism that counteracts gravity. Likewise, if the Universe is accelerating its expansion, presumably it needs some sort of anti-gravity, too.

Cosmologists who had done their homework realized immediately that general relativity allows for exactly such a term: the cosmological constant. Suddenly, Λ was no longer dead; it was feeling much better. Einstein had introduced Λ to fix a problem that didn't exist, but long after his death, it rose again, like a zombie, to explain the accelerating expansion of the Universe. Yet to

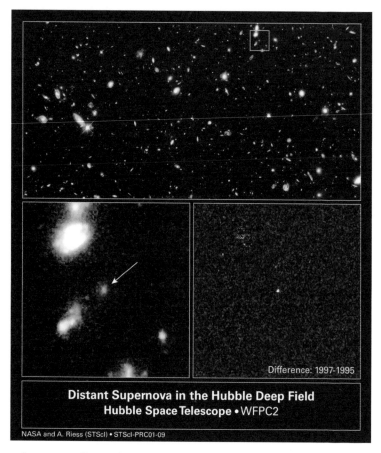

Distant Supernova in the Hubble Deep Field
Hubble Space Telescope • WFPC2

Difference: 1997-1995

8   Supernova 1997ff was, at the time, the most distant Type Ia supernova ever detected, seen 10 billion years ago. By using 1997ff's light curve as a standard candle, it was found that the Universe was decelerating at that epoch, in contrast with supernovae seen 6–7 billion years ago that indicated an accelerating Universe. The change between early deceleration and later acceleration is exactly what Einstein's cosmological constant predicts. sn1997ff ruled out many competing models for the dimness of Type Ia supernovae such as intergalactic dust, leaving dark energy as the favoured explanation.

many cosmologists, Λ seemed like mathemagics, a term added in for no rhyme or reason other than to make things work. It wasn't quite as bad as Einstein's invocation to balance a bowling ball on a pinhead, but it was still unsatisfying; surely, Λ must have some proper physical origin?

The most natural idea was that Λ, being a fundamental parameter of our Universe, was set at the Planck time when gravity broke off from the other three forces of nature. Perhaps something in the nature of the superforce gave rise to a cosmological constant. If so, it should have a value corresponding to the typical energy in the Universe at that time, the Planck energy. But it turns out that the Planck energy is 10^{120} times larger than the measured value of Λ! Yes, that number is 1 followed by 120 zeros. That's not

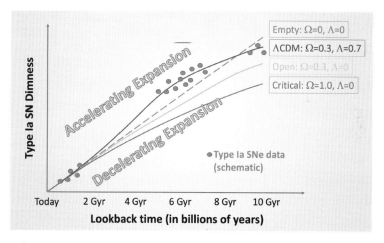

9 Type Ia supernovae imply an accelerating expansion rate. The supernovae seen 5–7 billion years ago are dimmer than expected as compared to an empty (constantly expanding) Universe, following expectations for a model currently dominated by dark energy. Beyond 8 billion years ago, supernovae are again brighter than in an empty Universe; back then, the Universe was decelerating because matter was compressed enough to dominate over dark energy. The data is best fitted by a model with 30 per cent of the critical density in matter (baryonic and dark) and 70 per cent in dark energy, which is concordant with the values inferred from baryon acoustic oscillations in the CMB, providing another pillar for the concordance ΛCDM cosmological model.

just a wrong prediction, that's an embarrassment. This is often called the worst prediction in the history of physics.

So theoretical cosmologists have been busily inventing alternatives to Einstein's simple cosmological constant. This is another Wild West area of cosmology today. Current alternatives include quintessence, Chaplygin gas, and extensions to general relativity called $f(R)$ gravity models that do away with Λ altogether. Yet stubbornly, all observations so far are best explained by Einstein's cosmological constant. It is hoped that the upcoming generation of telescopes such as the European Space Agency's Euclid satellite will find evidence that disfavours a cosmological constant and points towards the true nature of dark energy. But there is no guarantee of this; as with dark matter, the Universe is under no obligation to make all her secrets accessible to human technology.

COSMIC INFLATION: EVERYTHING FROM NOTHING

By the 1980s, the Big Bang model was mostly accepted because it nicely explained the CMB, and where hydrogen and helium came from in stars (BBN). But upon more careful inspection, it had problems. It was not that any observations contradicted the Big Bang, but rather that the standard Big Bang model required some remarkable coincidences to work. It is a very anti-Copernican idea that the Universe was somehow fine-tuned just for us; instead, if something appears fine-tuned, physicists usually take it as a sign of overlooked or misunderstood physics.

The first issue was that the CMB looked the same in every direction. That might not sound particularly problematic, but careful reflection reveals a quandary. Recall that the CMB has been streaming to us from almost 13.8 billion years ago. That means if we look in opposite directions, the two points we're seeing the CMB stream from are 27.6 billion light years away from each other. But if the Universe is only 13.8 billion years old, then those two points could never have been in causal contact. How did they know to have exactly the same temperature, fluctuation patterns and so on, down to the finest part-in-a-million detail? It's as surprising as if two people from opposite sides of the globe with no known relations somehow looked like identical twins, down to their eyebrow hairs. While not impossible, it certainly seems like a massive coincidence. In cosmology, this is known as the horizon problem.

Another issue was the flatness of the Universe. Why should it have exactly this Goldilocks critical mass-energy density? There was nothing in the Big Bang model that would prefer one value of the cosmic mass-energy density over any other, yet from among infinite possible values, the measured one turned out to be indistinguishably close to this magic critical density – another massive coincidence, dubbed the flatness problem.

A more esoteric issue came from models of particle physics that suggested when gravity split off from the other three fundamental forces at 10^{-43} seconds, a huge number of magnetic monopoles should be created. A magnetic monopole is an object that has a North (or South) pole, without a corresponding South (or North) pole. Such objects, evidently, do not exist in our Universe, as one of Maxwell's fundamental laws of electrodynamics tells us. So where did they go? This is known as the magnetic monopole problem.

In the early 1980s, two physicists on opposite sides of the Iron Curtain, Alan Guth at Stanford and Andrei Linde in the Soviet Union, independently postulated a similar single solution to all these issues: the Universe, at very early epochs, underwent a

sudden and extremely rapid phase of expansion. This idea came
to be called cosmic inflation.

Inflation would solve all three problems simultaneously. It
would solve the horizon problem by allowing the Universe to be
in causal contact prior to inflation, thus equilibrating to the same
temperature, before rapid expansion pushed those points out of
causal contact. It would solve the flatness problem because,
whatever random, curved geometry it had prior to inflation, the
rapid expansion would make our observable patch of the Universe
appear to be exceptionally flat afterwards – akin to standing on
a beach ball that suddenly blows up to the size of Earth. It would
solve the magnetic monopole problem by diluting the density
of magnetic monopoles so much that we would never expect to
encounter one.

The true beauty of inflation, however, and why it gained
widespread acceptance among cosmologists, is that it elegantly
explains the origin of those fluctuations that we see in the CMB.
Even more fascinatingly, inflation argues that it arises out of
(apparently) nothing. How is that possible?

You were probably taught that energy is always conserved,
and that a vacuum is the absence of matter. As with many things
we learn in school, these two notions are not precisely true when
dealing with the wacky world of quantum mechanics. According
to Heisenberg's uncertainty principle, it is possible to violate
energy conservation for a very brief time. Through $E = mc^2$, this
violated energy can convert into mass, creating a particle–antiparticle pair, so-called virtual particles, so long as they annihilate
again within a short time. So a vacuum isn't empty at all; it's full
of all these virtual particles. Today, such virtual particles are
unnoticeable, because the scale over which these quantum effects
occur is tiny, although the reality of such virtual particles can be

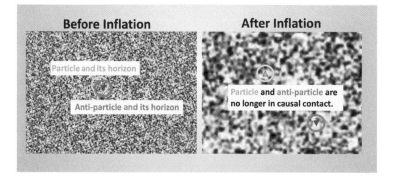

10 Inflation causes an extremely rapid expansion of the Universe. Particle–antiparticle pairs created just prior to inflation end up outside each other's horizons, so they can no longer annihilate. The matter is thus frozen, giving rise to the matter fluctuations that we later see in the cosmic microwave background.

demonstrated in the laboratory via the so-called Casimir effect. But just after the Big Bang, the entire Universe was tiny. Back then, these effects weren't small at all.

Imagine that at some point in space, a quantum fluctuation creates a particle–antiparticle pair. At that very instant, inflation happens. Now, just as the pair is about to annihilate again, they suddenly find themselves a huge distance apart, since space has expanded between them. They can't annihilate because they are no longer within each other's light horizon, so they cannot find each other (illus. 10). Those points in space now contain a bit of extra mass, and thus have ever-so-slightly higher densities. Not so virtual anymore, are they?

The energy for these created particles is extracted from the vacuum, which then loses energy. The vacuum must therefore contain some energy, called an inflaton field, which, as mentioned earlier, is thought to have come from the symmetry breaking of the GUT force. Through vacuum pair production, inflation converts this energy into mass, producing tiny mass fluctuations.

The crowning achievement of inflation is that it naturally predicts mass fluctuations that are a beautiful match to the mass fluctuations observed in the CMB at 380,000 years. This pattern is not like the 'white noise' one sees on an old-fashioned TV, but follows a characteristic shape known as a Harrison–Zeldovich spectrum. The fact that inflation elegantly yields a Harrison–Zeldovich spectrum of fluctuations out of basic quantum mechanics is a remarkable triumph of modern physics, and is seen as a strong point in its favour to this day, even in the face of many competing models. These inflation-generated fluctuations will be the basis for setting up our simulations of the cosmos.

COSMOLOGY IN CONCORDANCE

And so we arrive at the concordance cosmological model, which posits that we live in an accelerating, dark matter-dominated, inflationary Big Bang Universe. On the largest scales, it is excellently described by general relativity applied to a Universe that today contains:

(i) 5 per cent baryonic (ordinary) matter, with about three-quarters of the mass in hydrogen and a quarter in helium;
(ii) 25 per cent dark matter, which is some as-yet-unidentified subatomic particle that does not interact with photons;
(iii) 70 per cent dark energy, which is an unknown component that drives the accelerating expansion and is consistent with a vacuum pressure from a cosmological constant; and
(iv) A space-filling sea of photons (the CMB) streaming to us from 380,000 years after the Big Bang.

It is a remarkable achievement of scientific ingenuity and technological advancement that these numbers are now known to precisions of a few per cent, based on numerous complementary and independent measurements from the CMB and other sources. But in a way, it leaves more questions unanswered than answered, most notably the nature of dark matter and dark energy. Theoretical cosmologists continue to pursue these questions, and upcoming observatories will surely shed new light on the dark sector.

In the meantime, numerical cosmologists like myself are interested in a related but different question: why does the Universe look the way it does? When we say 'look', this implies stuff that we can see – that is, the 5 per cent of cosmic mass-energy in the form of baryonic matter that interacts with light. This is the 'stuff' that makes up the visible Universe, the galaxies, stars, black holes, interstellar gas clouds, dust, planets, moons and everything we can directly observe in our telescopes. We would like to understand why all of this forms within the concordance cosmological model, how it evolves over time, how it is distributed in space and what sets its properties. In the end, we want to connect the Big Bang to the formation of galaxies and stars and planets and eventually life, such as humans. This is nothing less than our new and improved cosmic origin story, now fully scientific and borne of observations and evidence. This is our charge.

To do this, we're going to need computers. Big ones.

2

PUTTING THE UNIVERSE ON A COMPUTER

The past century has taken us from viewing the cosmos as a static entity dominated at its centre by our Milky Way and the stars within it, to the notion of a dynamic Universe having a definite beginning, with our Milky Way being just one of billions of galaxies dotting its vastness. We've realized, in true Copernican fashion, that not only is Earth not special, but our Sun isn't special, our Galaxy isn't special and even the stuff we are made out of isn't special.

The human race stands on a precipice. An evidence-based cosmic origins story for humans would take what had once been in the realm of religious doctrine under the wing of science, to be questioned and investigated and debated towards a more complete understanding of how we and everything around us got here. The concordance cosmological model provides the mise en place for constructing an answer to the origin of everything we see.

How do we solve the grand mystery of why the Universe looks the way it does? Increasingly in the modern era, the answer is that we simulate it on a computer. By this, I mean that we put the relevant laws of physics into a computer, set up some initial conditions at an early cosmic epoch, add in all the ingredients we know of – such as baryonic (ordinary) matter, dark matter and dark energy – and let it all churn in the world's most powerful supercomputers until we produce a simulated universe of our very own.

If our simulated universe resembles the real Universe, we know we're on the right track. If it doesn't, it pushes us to investigate the physics that we might be missing, overlooking or getting wrong, and improve our simulations for the next go-round. Effectively, we utilize computer simulations as numerical experiments – in lieu of being able to conduct experiments on the real Universe – so that we can test our physical understanding of how the Universe came to look the way it does. This is the essence of numerical cosmology.

How do we do this in practice? How do we put the laws of physics into a computer? How do we shrink an entire Universe to fit inside it? In this section, we'll dig into the nuts and bolts of how cosmological simulations work.

PARTICLE PUSHERS

The basic idea of simulations is so simple that one wonders why it causes such a fuss. Fundamentally, the entire Universe is just a collection of particles interacting with their surroundings. Protons, neutrons and electrons interact to make atoms, which interact to make molecules, which interact to make substances, which interact to make objects or life or planets or whatnot, which in turn interact to make solar systems and galaxies, all the way up to the entire Universe. The laws of interaction are well known, and can describe the Universe all the way back to the Planck time of 10^{-43} seconds.

Simulating the cosmos then involves representing a chunk of the universe that contains a huge number of particles, and having them all interact via the known laws of physics. We start at the Big Bang, compute those interactions and determine how the particles should move or behave. We then move the particles a minute bit over some tiny step forward in time. If we keep repeating these timesteps until the simulation covers the 13.8 billion years that brought us to today, voila, there's our simulated universe! Easy enough, right?

The simplest form of a cosmological simulation is one that puts only the force of gravity into a computer. This is called an N-body simulation. Despite including only gravity, we can already learn a lot from N-body simulations, because the dominant mass component in the universe (dark matter) only responds to gravity.

Let's say we want to simulate a cubic patch of the universe. We do this by representing all the mass in that patch by particles. There are a large number of these particles (N 'bodies', where N is a large number), with each particle representing a chunk of mass. At the start, let's arrange these particles so that some regions of space are slightly denser – we do this by shifting those particles closer together. Other regions are slightly 'under-dense' compared to the average density because we place the particles in that region slightly further apart. We can do this precisely in a way that mimics the harmonic spectrum of density fluctuations which we measure from the CMB. This gives us our initial conditions, from which we are ready to begin our N-body simulation.

An N-body code is a computational engine that steps this system of N bodies forward in time (see illus. 11). To do this, at each timestep, the code must compute the gravitational force on each particle from every one of the other $N-1$ particles, using Newton's law of gravity (plus some minor relativistic modifications) – this is where we put the laws of physics into the computer. By summing up all the forces from all the other particles, this then gives a total force on a given particle. With the particle's mass m and the total force F, we can calculate the acceleration a on that particle using Newton's second law: $a = F / m$.

We use this acceleration to step this particle forward in time. We pick a timestep Δt. Since velocity is acceleration times time, over this timestep the particle's velocity will change by $\Delta v = a \times \Delta t$. We can then add this Δv to update the particle's velocity v.

Using this new velocity, we can analogously determine the distance a particle will move over that timestep. Since distance is velocity multiplied by time, the change in position will be $\Delta x = v \times \Delta t$. We add Δx to the particle's position, thereby moving the particle to its new location. This ends the timestep.

We do this procedure in each of the three Cartesian directions (x,y,z) independently. By using the force in each direction, we calculate the acceleration in each direction, which gives the change in velocity over the timestep Δt, which in turn gives us the change in the particle's position in each direction. By the end of the timestep, we have updated the particle's (x,y,z) position and velocity to a slightly later time.

After this, it's lather, rinse, repeat. We keep doing this exact same procedure, stepping forward by Δt in time over each timestep, until we reach 13.8 billion years after the Big Bang. This can take thousands or even millions of timesteps; fortunately, computers are extremely fast and accurate at repetitive procedures. By the end, all the particles have now moved from their initial positions, following the law of gravity, to their final position. This represents the distribution of mass in the Universe in our simulated cube today.

Look at that . . . we have just simulated the cosmos!

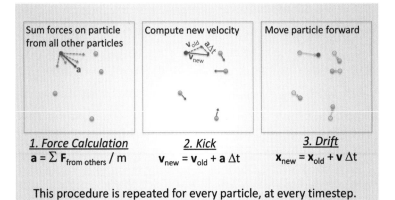

Sum forces on particle from all other particles	Compute new velocity	Move particle forward
1. Force Calculation $a = \sum F_{\text{from others}} / m$	**2. Kick** $v_{new} = v_{old} + a\,\Delta t$	**3. Drift** $x_{new} = x_{old} + v\,\Delta t$

This procedure is repeated for every particle, at every timestep.

11 Schematic of a single timestep of an *N*-body simulation. A system of *N* particles (where $N = 5$ here) begins with some initial positions and velocities. In the force calculation step, the gravitational force on the red particle from every other particle is calculated (green dashed arrows) and summed, which when divided by the red particle's mass, gives the acceleration for the red particle (red arrow). This is repeated for every particle. In the kick step, the acceleration is used to change the velocity of the particle over a timestep Δt. Finally, in the drift step, the velocity is used to move the particles forward in time by Δt, from the old (lightly shaded) positions to the new ones. This sets the stage for the next timestep. Such timesteps are repeated until the simulated universe has reached an age of 13.8 billion years.

GETTING RID OF COSMIC EXPANSION

The timestepping procedure described above used Newton's law of gravity, and didn't mention anything about expanding space-time. Shouldn't it be necessary to include some general relativity, or other theory, in order to account for cosmic expansion?

The answer is yes, but luckily this turns out to be straight-forward. The trick is choosing the right coordinate system in which to do the calculation, so we make things appear (mostly) Newtonian in our equations. To do so, we will need to introduce a correction to the Newtonian motion that accounts for cosmic expansion.

The coordinate system that turns out to be most convenient is one that expands with the Universe. Going way back to illustration 1, it's the growing coordinate grid for the 'expansion of space' case. In cosmology, this is known as co-moving coordinates.

Co-moving coordinates can be thought of as the rest frame of the CMB. If one is moving exactly with co-moving coordinates, the CMB will be isotropic (the same in every direction). However, the effects of local gravitational attractions can cause departures from this. For instance, here on Earth, we are moving with respect to co-moving coordinates because we are orbiting the Sun, the Sun is orbiting within the Milky Way and the Milky Way is being tugged by structures around it, for example, the Andromeda Galaxy and the Great Attractor. This gives the Earth what astronomers call a peculiar velocity.

Peculiar velocity can either be thought of as our motion with respect to co-moving coordinates, or equivalently our motion with respect to the CMB. It is called peculiar because it is unique to each object. Our peculiar velocity makes us feel a CMB 'wind', which we can measure as a Doppler shift of CMB photons. Currently, the Sun has a peculiar velocity of about 370 km/s; this will change over time as our Sun orbits around our Galaxy.

The tricky bit is that co-moving coordinates are non-inertial. By this, I mean that an object at rest in co-moving coordinates would be accelerating with respect to a fixed coordinate system. The trouble with non-inertial frames is that they introduce fictitious forces. 'Fictitious forces' sounds bizarre, but you've surely experienced them yourself. For instance, as an elevator begins moving upwards you feel a force pushing you downwards; this is a fictitious force, where suddenly you feel a force coming from seemingly nowhere because your coordinate system inside the elevator is being accelerated. Similarly, when a train hurtles

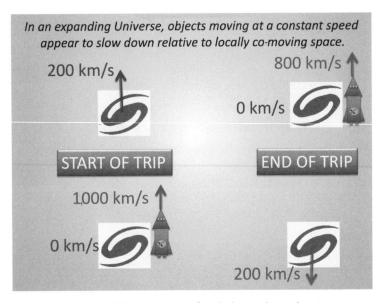

In an expanding Universe, objects moving at a constant speed appear to slow down relative to locally co-moving space.

200 km/s

800 km/s

0 km/s

START OF TRIP

END OF TRIP

1,000 km/s

0 km/s

200 km/s

12   Hubble drag illustrated for a rocket journey from the lower galaxy to the upper one, with both galaxies assumed to be stationary in co-moving coordinates. Initially, the rocket's peculiar velocity (relative to co-moving coordinates) is 1,000 km/s. But when it arrives at the location of the other galaxy, the latter has already been receding at 200 km/s, so now the rocket's peculiar velocity at its new location is only 800 km/s. This shows how the non-inertial co-moving coordinate system imparts a 'drag' on the rocket and appears to slow it down, relative to its surroundings. In actuality, the rocket is moving at the same speed (relative to a fixed coordinate system), but co-moving coordinates have effectively absorbed some of its velocity owing to cosmic expansion.

around a bend or you ride on a merry-go-round, you feel yourself pushed to one side; this is a fictitious force known as centrifugal force. A fictitious force always occurs when you are computing the laws of physics in a coordinate system that is accelerating with respect to a fixed coordinate system. Despite being called fictitious, the force is very real to the person in the elevator or on the merry-go-round.

Co-moving coordinates are an accelerating frame and thus they introduce a fictitious force called Hubble drag. Hubble drag can be thought of as an apparent slowing down of your peculiar velocity once you begin to move with respect to the co-moving coordinates.

Say there are two galaxies moving apart at 200 km/s, both moving with co-moving coordinates (that is, each is stationary with respect to the CMB). In one galaxy, Elon Musk Jr sets off in a rocket ship at 1,000 km/s towards the other galaxy. At the start, Elon Jr measures his peculiar velocity (that is, his velocity relative to co-moving coordinates) to be 1,000 km/s. After a short time, the rocket reaches the other galaxy. But since that galaxy was already receding at 200 km/s, Elon Jr now measures his peculiar velocity to be only 800 km/s (illus. 12).

Elon Jr has just experienced Hubble drag. Even though his rocket is always moving at the same speed, when measured in co-moving coordinates, he has appeared to slow down. This deceleration is caused by a fictitious force introduced by using co-moving coordinates.

Fortunately, Hubble drag is fairly simple to include in the equations for evolving a particle forward in time – the speed simply slows by the difference in the co-moving velocities of the initial and final positions. This correction accounts for the general relativistic effects of cosmic expansion during the timestepping of our particles. This is also where dark energy factors in to the simulation, since the expansion rate (and hence the co-moving velocities) depends on the cosmological constant Λ. This is how variations in global cosmological parameters impact the evolution of particles in a simulation.

THE DYNAMIC RANGE PROBLEM

So far, this all seems pretty easy. But before we get too smug, let's hear the bad news. I'll begin with a simple calculation. Don't worry, it won't get too technical, even if the numbers are, well, astronomical.

The Universe is made up of galaxies, and galaxies are made up of stars. Stars themselves are made up of hydrogen and helium atoms (along with some other stuff, but let's ignore that). Let's consider each star as a single 'particle' in our N-body simulation. How many particles do we need to represent the Observable Universe?

The Observable Universe has a radius of about 45 billion light years; this is the present-day distance to objects that were 13.8 billion light years away 13.8 billion years ago, and hence the distance from which light has had time to reach us since the Big Bang. Using the formula for the volume of a sphere, this means that the volume of the Observable Universe is about 10^{13} (10 trillion) cubic megaparsecs (Mpc). Meanwhile, the average separation between galaxies like the Milky Way is about 1 Mpc. This means that, within the volume of the Observable Universe, there are of the order of 10^{13} Milky Ways. Each Milky Way contains, conservatively, about 100 billion stars. Putting it together, this means that modelling the entire Observable Universe with stars as 'particles' would require modelling about 10^{24} stars – a million exastars – on a computer. If that seems like a big number, you're right.

To simulate each star, we will need to store some basic information about its properties in the computer. At minimum, one would need the mass, position, velocity and acceleration of each star. The mass is a single number, but the other quantities are all three-dimensional, so we need ten numbers to store in the computer's memory for each star. In a computer, a number takes up 8 bytes of memory, so that is about 100 bytes per star, give or take. To store all our stars in the Observable Universe, we are talking about 100 million exabytes (10^{26} bytes) of memory, in a single computer. To come to grips with how much memory this is, if we added up all the memory in all the computers in the entire world today, it would come to less than 10,000 exabytes.

So even just to fit all the stars in the Observable Universe on a computer, let alone any gas or dark matter or anything else, we would far exceed the total storage capability of all the computers in existence today. We haven't even begun to compute anything yet, and we've already blown well past the entire world's computing budget!

But it gets worse. There's also a matter of resolution. To see what I mean, consider my MacBook. It has a screen resolution of $3{,}024 \times 1{,}964$ pixels, high enough so that my eyes don't notice any pixelization. Nonetheless it does limit what I can display on my screen, because the number of pixels is not infinite.

Numerical cosmologists face an analogous problem. Let's say we want to locate the Sun's position in our Galaxy. Our Sun is about 25,000 light years away from the centre of the Milky Way. How many pixels do we need in our simulation of the Observable Universe (90 million light years across)? The answer is, we would need about 36 million pixels along each axis. I don't think Apple is going to come out with that kind of screen resolution anytime soon.

But it gets worse. Since the Universe is a three-dimensional object, we need the same number of pixels in each direction. In other words, the number of three-dimensional pixels (or voxels) we need is more than 36 million cubed, or about 5 million billion mega-voxels. That number is mind-boggling. And even if we get that, it's only representing the entire Milky Way by a few blocky pixels, Minecraft-style; it's nowhere near close to being able to resolve individual stars.

And it gets worse again. Remember, each of these 5 billion mega-voxels elements has to compute its gravitational interaction with every one of the 5 billion mega-voxels. So that is over 10^{45} calculations to evaluate the forces. The fastest processor today would take more than 10^{29} seconds, or over 100,000 times the age of the Universe, to compute these forces. There doesn't seem much point in simulating the cosmos if the simulation is going to take far longer to run than the cosmos has been around.

But wait . . . you guessed it . . . it gets worse. Remember, this is just the force calculation for a single timestep. We have to do thousands or millions of timesteps to evolve our simulated universe over the 13.8 billion years since the Big Bang. Besides the sheer number of computations, the number of timesteps also scales with the resolution. This is because we can't take too large a timestep, or else our particles would jump all over our simulation volume willy-nilly, and would not accurately respond to the forces from the other particles around them. In other words, there is effectively another dimension that we have to resolve as well, namely time.

As a result, the number of calculations needed to run a cosmological simulation scales with the seventh power of the number of particles: one for each dimension to represent particles (3), each interacting with every other particle (3 × 2), plus a dimension for time (3 × 2 + 1 = 7). What if we want to resolve that Milky Way with twice as much resolution as you had before? Sure, no problem, just find a computer that is $2^7 = 128$ times as powerful! With that kind of scaling, it's easy to see how the problem of simulating the cosmos becomes intractable very quickly.

This is known as the dynamic range problem. Dynamic range is the ratio of the largest scale we want to represent (that is, the size of the Observable Universe) to the smallest scale (that is, the typical distance between stars). As Douglas Adams so succinctly put it, space is big . . . really big. It's way too big to fit into the biggest computers we have today, or will have in the foreseeable future.

So now what? Are we numerical cosmologists going to throw up our hands and quit? Heck, no! We will simply have to make some compromises. It turns out we can still learn a whole lot from simulations even if we can't fit all the stars in the Observable Universe into a computer at once. Here are some clever approaches used to sidestep the dynamic range problem:

1   We reduce the volume we want to model. Maybe we don't need to do the entire Observable Universe, maybe we only need to model a representative portion of the entire Universe.

2   We make the simulation coarser. Maybe we can't model single stars, but what about groups of stars? We could have each particle correspond to thousands or millions of stars' worth of mass. In doing so, obviously we lose information about specific stars like the Sun, but we can quickly reduce the number of particles that we need to store and compute.

3   We find ways to compute gravitational forces faster using approximations and computer science algorithms, so that the computational time doesn't scale with the seventh power of dynamic range.

4   We do all of the above.

The right answer is, naturally, all of the above. Let's see how we can implement these time-saving tricks.

THE BUTTERFLY AND THE MEADOW

Let's discuss the first approach. Suppose, rather than trying to do the entire Universe at once, we break up the problem into manageable chunks of dynamic range. A star such as the Sun going around in the Milky Way doesn't feel cosmic expansion, so it doesn't care much about what's going on halfway across the Universe. As such, we don't need to simultaneously model a single star in a galaxy while also modelling faraway galaxies. Maybe we can try to assemble the story from looking at various simulations covering different scales of the problem.

This is analogous to an ecologist trying to study an individual butterfly species along with the variegated flora in a large meadow. On large scales, the ecologist takes a photograph of the flowers in the meadow, and then close-up shots of individual butter-flies and flowers. From these, the ecologist pieces together the story of how each butterfly is adapted to its overall environment. Numerical cosmologists take essentially the same approach, running coarse simulations covering large volumes alongside high-resolution simulations covering small volumes.

This gives rise to the idea of tiered simulations. The coarsest tier, Hubble volume simulations, spans billions of light years in a single simulation. The resolution, however, is poor; the entire Milky Way would fit into a single pixel, but the matter distribution of very large scales is accurately represented. On the next tier are so-called cosmological simulations, which are typically done in cubic volumes of hundreds of millions of light years along each dimension of the cube, so they can resolve the locations and properties of individual galaxies, yet still have many thousands or millions of galaxies within a single volume. Finally, the finest-level tier zoom simulations simulate individual galaxies in great detail, while sacrificing statistics.

Examples of recent simulations at these three tiers are shown in illustration 13. From left to right, we can view a projection of the matter in a Hubble volume simulation called Millennium XXL, a cosmological simulation called EAGLE and a zoom simulation from the FIRE project. The increased level of detail at each tier is evident. Each of these simulations took weeks to months to run on state-of-the-art supercomputers, so they are pushing the limits of current technology. By combining the information from these simulations covering different scales, numerical cosmologists can effectively span a much wider dynamic range than a single simulation that tries to do everything at once, for a manageable computational cost.

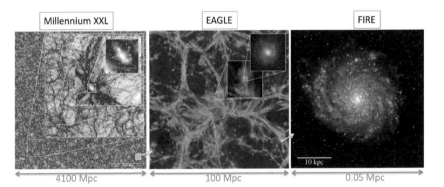

13 Tiers of cosmological simulations. The left image is from a Hubble volume simulation called Millennium XXL, the middle image is a cosmological simulation from Durham University's Institute for Computational Cosmology called EAGLE and the right image is a zoom simulation from Northwestern University's FIRE project. The small yellow box in the lower right of Millennium XXL shows a comparable volume to EAGLE, while the tiniest box in EAGLE shows the size of the FIRE image.

An advantage of running various tiers of simulations is the ability to do resolution convergence tests. We can look at the detailed structure inside a galaxy in our zoom simulations and ask, how well is this represented in our cosmological simulation with poorer resolution? By quantifying the discrepancies among tiers, we can assign an uncertainty to the results from our simulations, which is necessary when trying to do proper statistical analyses. Resolution convergence tests are an important way to gauge what information from the simulations can be trusted.

In the end, each simulation must be tailored to answer a specific set of science questions, and many choices must be made to optimize the accuracy and believability within the available computing resources. This is what makes doing numerical cosmology not just a science, but something of an art.

ACCELERATING THE ACCELERATION

Another approach to overcoming the dynamic range problem is to speed up the force computation. Clearly, if we can speed this up, then within a given amount of computing time we can calculate more particles. This veers into computer science, by taking advantage of clever data structures, parallel and multi-threaded algorithms, and optimizations tailored to cutting-edge hardware.

The timestep schematic in illustration 11 depicted the simplest and slowest way to compute gravity. Each particle has N^{-1} forces to compute from every other particle, and this must be done for each of N particles. This requires $N \times (N^{-1})$ computations, which for large numbers is essentially N^2. If one wants to have a simulation size that has twice as many particles in each direction, this means eight times the volume, which means $2^3 = 8$ times the particles, which means $8^2 = 64$ times the number of calculations. Such N^2 scaling quickly becomes problematic.

But what if we could make it so that eight times as many particles only needed, say, ten times as many calculations, instead of 64 times? This would make bigger simulations a lot faster, or alternatively, for a given amount of computing power it becomes possible to have a larger dynamic range.

To do this requires a willingness to live with a manageable level of inaccuracy. There are two main algorithms used to speed up the force calculation:

1   Particle-Mesh (PM): Compute the forces on a uniform grid.

2   Tree: Compute the forces directly on the particles, using a data structure known as a tree.

Both methods reduce the scaling of the force calculation time with particle number from N^2 to $N\log_2 N$. For example, if a simulation has a million particles ($N = 10^6$), then N^2 would be a trillion (10^{12}), but $N\log_2 N$ is only about 20 million (2×10^7). You can see that for this simulation, we save an enormous factor in computation time – roughly 50,000 times faster with $N\log_2 N$ scaling in this case. Indeed, this is what makes modern cosmological simulations feasible. Let's see how these two approaches work.

PM Code: in a Particle-Mesh (PM) simulation (illus. 14), the particles are first smoothed onto the uniform grid, and the mass density is computed within each grid cell. This is where clever maths comes in: the grid of mass densities is transformed into Fourier space, which may be familiar to those with an engineering background as a representation of the field using a sum of harmonics (that is, sine curves) of different frequencies. In Fourier space, it turns out that the law of gravity can be represented as a simple multiplication. In other words, to compute the force of gravity in each cell requires only a single multiplication, instead of a multiplication involving every other cell. This now scales as N instead of N^2. The forces can thus be computed very quickly – albeit in Fourier space, but this can be reverted back to real space with an inverse Fourier transform. Finally, the grid forces are interpolated back from the grid to the particle positions to get the force on each particle.

PM codes are extremely fast. The multiplication required to get the force of gravity is basically free. The smoothing of particles onto a grid and interpolating back from the grid to get the force on each particle is not trivial, but it still scales directly with the number of particles N. The hard bit is computing the Fourier transform, but fortunately computer scientists have invented an algorithm known as a fast Fourier transform (FFT), which scales with the number of grid cells as $N\log_2 N$. The first big cosmological simulations in the mid-1980s were run using a PM code.

However, PM comes with a huge drawback: it is not adaptive. This means that when mass collects into a small region, it eventually becomes confined to one or few grid cells, and all the details below the size of the grid cell are lost (as in the right-side image of illus. 14). This is problematic, since we are often most interested in dense regions (which host galaxies) – and these are exactly the regions that are poorly represented in a PM code. Because of this, PM codes are ideal for representing the cosmos on large scales, but not optimal for studying individual objects on small scales.

An alternative algorithm is a tree code, which aims to preserve adaptivity while retaining the same beneficial $N\log_2 N$ scaling as

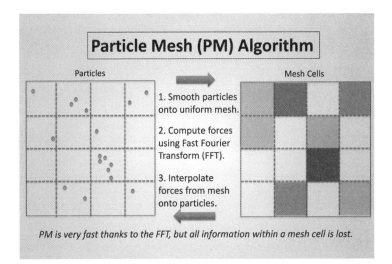

14 Particle-mesh algorithm, showing the particle distribution on the left, which is smoothed into a mesh density field on the right; then a Fourier transform is taken and Poisson's equation is solved on the density field to get the force in Fourier space. An inverse Fourier transform is applied to get the force back in physical space, which can then be interpolated back onto the particles. Because Fourier transforms on uniform grids can be done extremely quickly on computers, this algorithm is highly efficient, at the cost of losing all information about the particle distribution within cells.

a PM code. In computer science, a tree doesn't refer to a large bark-covered arboraceous growth, but instead to a particular type of data structure into which the particles are organized, which allows for computation of the forces. Let's see how this works. Warning: it's conceptually more difficult than a PM code, but we'll still discuss it because it's such a nifty application of computer science data structures to solve a physics problem.

First, the algorithm builds a tree structure from a set of particles. To construct a tree, one starts at the top (root) node, which covers the entire simulation volume and contains all the particles. This is subdivided into eight equal sub-volumes. If a sub-volume has more than one particle, it is subdivided again into eight smaller

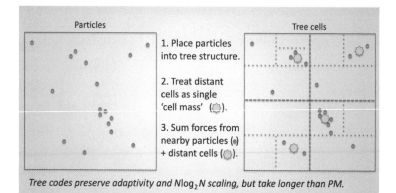

15 A tree code. From an overall particle distribution, a tree is constructed with each level of the tree representing a subdivision of space. Nearby particles have their force computed exactly, but more distant collections of particles are treated collectively based on their centres of mass (denoted by stars). This results in substantially fewer calculations needed to compute the force, and mostly retains $N\log_2 N$ scaling. The level of inaccuracy can be chosen via the so-called opening angle.

sub-sub-volumes, and so on, until each subdivided cell contains a single particle. Illustration 15 shows this in two dimensions. One can view this as branches of a tree, with (up to) four branches from each stem (which would be eight branches in three dimensions).

How can a tree be used to speed up the gravity calculation? The basic idea is this: since the force of gravity drops off as distance squared, this means that mass that is very close to a given particle has a large impact, while distant mass has a small impact. If we don't mind sacrificing a bit of accuracy, we can approximate the force from a clump of distant particles as coming from one giant 'cell mass' located at the centre of mass of those distant particles. If a distant cell contains six particles (as in the densest cell in illus. 15), and is far enough away to consider as a single cell mass, then we've just saved a factor of six in computing time for that force calculation over the naive N^2 approach.

The trick is, we must be able to figure out, for any given particle, which other particles are nearby enough that their force must be computed exactly, and which (groups of) particles are sufficiently distant that they can be treated as a single pseudo-particle. This is where the tree data structure comes in. In a tree walk, for each particle starting at the root node, one computes the distance between a given particle and the tree cell. If the cell is close by, we open the cell down to the next level of sub-cells. If a cell is sufficiently distant, however, we can treat all the particles within that cell as a single cell mass. As such each particle, instead of interacting with N other particles, only has to interact with roughly $\log_2 N$ other particles/cells. This gives us the desired $N\log_2 N$ scaling.

While the PM and tree methods both scale with the number of particles as $N\log_2 N$, the tree code is much slower because of the overhead for tree construction and tree walking. As such, the most popular method today involves combining these two to get the best of both worlds: the PM method is used on large scales, but in those few PM cells where lots of particles collect, the forces within those cells are computed using a tree. This allows us to keep computing forces accurately down to well below the PM cell size, thereby gaining the adaptivity advantage of a tree code, while having much of the universe's (relatively empty) volume computed using the faster PM code. Today, the PM-Tree method is the most common approach used when calculating gravitational forces in simulations.

A KINDER, GENTLER GRAVITY

There's an old adage that numerical simulations are like quantum mechanics: anything that is not explicitly forbidden will eventually happen, and even things that are explicitly forbidden will happen, only less frequently. As such, simulators must take care to examine all possible eventualities, however unlikely, and make sure that none of them will lead to unwanted catastrophic behaviour.

One example of catastrophic behaviour is when the distance between particles goes towards zero. Why is this a problem? You might recall that the force of gravity between particles of mass M_1 and M_2 at a distance R apart is

$$F_{\text{gravity}} = \frac{G\, M_1\, M_2}{R^2}$$

Without getting into the details, the point is this: if R goes to zero, then you can see this results in a divide-by-zero. This leads to F_{gravity} becoming undefined. This isn't realistic; it's an artefact of machine precision that the computer's binary representation for R ends up being identical for two particles. While R being exactly zero is rare, even for small R the force becomes unrealistically large. How can we handle this R-going-to-zero situation more realistically?

The solution is to soften the force of gravity. When particles get very close to each other, we put a cap on how large the force can get. We choose a softening length ε and calculate a 'softened' gravitational force as follows:

$$F_{\text{softened}} = \frac{G\, M_1\, M_2}{(R + \varepsilon)^2}$$

You can see now that if R goes to zero, instead of the force becoming undefined, the softened force F_{softened} only ever gets to a maximum value equal to $G\, M_1\, M_2 / \varepsilon^2$. With this force softening included, even particles that happen to come very close to each other behave nicely (see illus. 16) and the kingdom is saved.

As with any approximation, there is a cost: the force of gravity is inaccurately computed when particles are too close. In effect, force softening sets the spatial resolution, that is, the minimum length scale that can be faithfully represented in a given simulation. Clearly, length scales smaller than the softening length ε cannot be faithfully represented, since gravity is not being computed accurately there.

With a softening length, we can now go back and more formally define a concept that was discussed earlier: dynamic range. Recall that dynamic range is the ratio of the largest size to smallest size that can be represented within a single simulation. A simulation's box length L is the largest scale that can be represented, while the softening length ε sets the minimum scale that can be represented. Hence the dynamic range is defined as L / ε.

As an example, among the largest N-body simulations today is the Euclid Flagship simulation, for which $L = 5.4$ billion parsecs (pc) and $\varepsilon = 5,000$ pc, making its dynamic range a bit over a million. This is quite impressive; it's effectively like having a view of the Universe with a million pixels in each of the three dimensions. But it's still far short of what what we ideally need in order to represent the Universe.

AVOIDING THE SOUTH PARK EFFECT

A cosmological simulation must cover most of the age of the Universe, almost 13.8 billion years. The larger the timestep, the fewer the timesteps that will be needed to cover that time. Since the number of force calculations to run a full simulation scales as the number of timesteps, fewer timesteps means less computing time.

Faster is better, of course, but as usual there's a trade-off. Larger timesteps also mean that particles are literally jumping all over the place, and the animation becomes jumpy itself – think of the animated show *South Park* (1997–). Ideally, we want the timesteps to be small enough to represent a sufficiently smooth motion, but not any smaller than necessary to avoid wasting computing time. How large a timestep can we get away with?

The softening length ε gives us a clue for this. Since there is no reliable information in the simulation on length scales less than ε, we can perhaps get away with moving particles by ε in a single timestep. In practice, to get sufficiently smooth motion requires moving perhaps only 0.2ε in a single step. If the particle speed is v, then its timestep should be

$$\Delta t = \frac{0.2\ \varepsilon}{v}$$

But hang on . . . what happens when v equals zero? Since anything divided by zero is infinite, this means if v is small the timestep Δt will become very large. Why is this a problem? Imagine simulating a ball thrown in the air. At the top of the flight, the velocity will be zero, and the timestep infinite. As a result, our simulation predicts that the ball will sit there at the top of its arc forever. This is obviously wrong.

The way around this is for the timestep criterion to account for not only the velocity, but the acceleration. Even if the velocity is zero, if the acceleration is non-zero, then the timestep should still be finite. In our thrown-ball case, Earth's acceleration is still present even though the ball's velocity is zero.

To account for this, simulators add a second timestep criterion based on the acceleration: $\Delta t = 0.2\ \sqrt{(\varepsilon/a)}$, where a is the magnitude of the particle's acceleration. So in the end, the timestep for any given particle is then the smaller of the velocity and acceleration criteria:

$$\Delta t = 0.2 \, \text{MIN}[\ \varepsilon/v, \sqrt{(\varepsilon/a)}\ ]$$

Here, MIN means take the minimum value among those two numbers. This way we ensure the motion of the particles is always smooth on the spatial scales that we can resolve.

How can we implement this in a simulation? Well, it's tricky, because each particle has its own v and its own a, thus its own Δt. The conservative approach would be to find the minimum timestep among all particles in the entire simulation, and use that for all particles. That's safe, but also needlessly expensive – we would be computing forces too often for a lot of particles on large timesteps just to accommodate a handful of particles with the smallest timesteps.

To avoid this, simulations use adaptive timestepping. Here, each particle is put into a 'timestep bin' such that its forces are only computed when needed, alongside other particles that

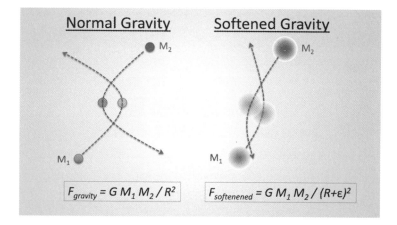

16 Without softening (left column – that is, normal gravity) the two point-like particles get close to each other and suffer a large deflection with high velocity, which is not physically correct. With softened gravity (right column), the particles are thought of as fuzzy 'blobs' of matter, and when they get within each other's softening zone ($r < \varepsilon$), the gravitational force is reduced, which causes a less extreme deflection. In *N*-body simulations, particles typically represent agglomerations of matter, so softened gravity is more realistic.

have a similar timestep. As particles change their velocity and acceleration during a simulation, they are shifted between time-step bins as necessary. This requires a lot of bookkeeping in the code, but it's not difficult in principle. In this way, the simulation minimizes the number of force calculations needed in order to keep the motion smooth. Combined with a tree code, this makes our simulation adaptive in both space and time.

THE STARTING LINE

The first rule of computing is, never use a computer unless you absolutely need to. Computers are slow, expensive, annoyingly finicky and intrinsically have round-off errors. In comparison, solving problems with a pencil is cheap, fast and has no round-off errors. Anything that can possibly be solved by hand should be solved by hand.

The early Universe is solvable with a pencil and paper. It involves some complicated physics, but luckily most of it was solved decades ago (back when there were no computers). Our simulation only needs to begin once the pencil-and-paper approach becomes hopelessly inaccurate. When is that?

The CMB gives us an image of the baby Universe when it was 380,000 years old. At that time, the measured fluctuations in the matter density represent deviations of just a few parts per million, relative to the average density. This type of situation where deviations from the average are tiny is amenable to a technique known as linear perturbation theory. This was applied to small matter fluctuations in the 1960s and '70s, most notably by Russian cosmologist Yakov Zeldovich.

But gravitational instability is an exponentially cruel mistress. It takes tiny perturbations and grows them exponentially fast.

It's like the old parable about rice grains on a chessboard – starting with a single grain of rice and doubling it over each of the 64 chess squares requires more rice than exists in all the world. That's analogous to how exponential growth takes those very tiny perturbations and transforms them, surprisingly fast, into fluctuations whose amplitude becomes comparable to average density itself. At this point, the Universe becomes non-linear – that is, the deviations from the average density become larger than the average density itself. Linear perturbation theory cannot accurately describe non-linear growth. This is when we must fire up the computer.

Non-linearity typically happens at around 100 million years, depending somewhat on the box size and resolution. So we can measure the mass fluctuations from the CMB at 380,000 years and evolve them forward using linear perturbation theory to 100 million years. At this point, we need to lay down our simulation particles.

To represent the matter fluctuations, we slightly displace the positions of the particles from grid points in just such a way that represents the desired density variations. In other words, in some regions, we place the particles slightly closer to each other to represent a slightly over-dense region, while in other places we move the particles a bit further apart to represent an under-dense region. We also must give them some velocity, since particles move towards dense regions and away from voids. This procedure is known as the Zeldovich approximation. By choosing our displacements and velocities this way, we can (statistically) represent the mass fluctuations occurring in the real Universe at 100 million years. This gives us the initial conditions from which we will start timestepping our simulation forwards.

AND WE'RE OFF!

To sum it up, here are the parameters we have to choose to set up a simulation of the Universe:

Cosmology: We must choose values for the mass densities of dark matter (Ω_m) and dark energy (Ω_Λ) in our Universe, the cosmic expansion rate (usually scaled by today's Hubble constant H_0) and any other specific cosmological parameters. These parameters will set the total mass to be simulated within a given volume, and how the expansion rate (for example, Hubble drag) will affect the motions of particles.

Box size: Typically, simulations are done in a cube. The parameter that determines the volume is the box size L – the length of a side of the cube. The simulation volume is L^3.

Particle number N: The given mass within our volume is represented by N particles. With higher N, each particle has less mass, so N effectively sets the mass of each particle. A higher N is desirable, but it's limited by the available computing resources.

Softening length ε: This sets the minimum spatial scale that can be reliably represented in the simulation, and thus sets the dynamic range (L/ε). Smaller is better, but it can't be too small or we risk having particles flying around artificially fast.

Timestep: We must choose a pre-factor applied to the minimum of ε/v and $\sqrt{(\varepsilon/a)}$, where v and a are the magnitude of the particle's velocity and acceleration, respectively. A pre-factor of 0.2 is typically found to represent smooth-enough motion for most applications.

Starting time: This is set so the chosen volume is just entering the non-linear regime, typically around 50–100 million years after the Big Bang. This is when the initial conditions must be generated that mimic the mass fluctuations observed in the CMB.

And that's it! These are all the ingredients we need to run a cosmological N-body (gravity-only) simulation.

These days, there are freely available state-of-the-art computer programs that one can use to set up and evolve cosmological simulations. At present the most popular such package is the so-called Gadget code, written by Volker Springel, the head of the Max Planck Institute for Astrophysics, in Germany. The latest version, Gadget-4, contains all the codes required to both set up and run a simulation on a parallel supercomputer, and even do some basic analysis on the outputs. A new code called SWIFT, developed by the good folks at Durham University's Institute for Computational Cosmology, is also publicly available, and offers faster speeds than Gadget-4 but as-yet less developed analysis toolkits. Indeed, these and various other codes, such as Enzo-E, RAMSES and AREPO, also include extensions for hydrodynamics to simulate the formation of galaxies, as we will discuss later. The codes are written in the programming languages C/C++, and are mostly tailored to a Unix/Linux environment. With these, you too can run your own cosmological simulation. It helps to have a supercomputer lying around, but the progress of technology is such that the simulations I ran for my PhD thesis on massive parallel supercomputers in the late 1990s can today be run on any good gaming server.

Hopefully you now have a better appreciation of the nuts and bolts of how N-body simulations work. But what can we learn from them? The answer is, a huge amount. As an example, simulations played a critical role in establishing the aforementioned concordance cosmological model, and in essence fundamentally altered the way that cosmology was studied. Let's see how this came about.

3

THE SIMULATED UNIVERSE: HALOS IN A WEB

In the 1980s, following in the footsteps of the ancient cartographers, observational cosmologists began mapping out the Universe around us. This was achieved using redshift surveys. A redshift survey is a three-dimensional map of galaxies, in which the two-dimensional positions in the sky are determined from images, while the distance to each galaxy is estimated by obtaining its spectrum and measuring the redshift, or Doppler shift, of absorption or emission features. The Doppler shift tells you the recession velocity, which, using the Hubble–Lemaître Law, gives the distance. It is only an approximate distance because peculiar velocities can cause slight deviations from the Hubble–Lemaître Law, but when looking out to large distances the inaccuracies become small.

The first redshift survey began in the mid-1980s, with the Harvard–Smithsonian Center for Astrophysics (cfA) Redshift Survey. This caused quite a stir. Astronomers were surprised to discover that galaxies in the sky were not distributed randomly, but instead were organized into a complex network of filaments, sheets and nodes.

An example of such a redshift survey is shown in illustration 17, which depicts the galaxy distribution in the Two Degree Field (2dF) Survey, completed using the Anglo-Australian Telescope in the late 1990s. It is apparent that galaxies are arranged into connected structures like filaments and sheets, whose intersections at the nodes host 'cities' of galaxies called galaxy clusters. This large-scale structure came to be known as the cosmic web.

The cosmic web was an eye-opener for cosmologists. Surely this pattern had some great and fundamental significance. Questions immediately began to bubble forth: what causes galaxies to arrange themselves in this way? Does this pattern evolve from the early Universe until now? How does this pattern depend upon the cosmological parameters of the Universe? And most importantly, how can we use the cosmic web to learn about cosmology?

SIMULATIONS VERSUS SURVEYS

As it happened, right around the time of the first redshift surveys in the 1980s, the first cosmological N-body simulations were being run. Seminal work in 1984 by George Blumenthal, Joel Primack, Marc Davis and Sandra Faber presented pioneering N-body simulations of a Cold Dark Matter universe. Remarkably, these very first simulations showed that the distribution of matter in the Universe was not uniform, but rather arranged into filaments, sheets and nodes, just as was seen in redshift surveys.

This was considered a great triumph for these early simulations, and a remarkable confirmation for the inflationary Big Bang scenario. It helped greatly in gaining acceptance of cosmological simulations within the wider astrophysical community, countering some of the natural scepticism that follows the introduction of a new technique. However, as numerical cosmologists examined their simulations more closely, it became clear that the detailed patterns of the filaments, sheets and nodes in the cosmic web predicted by these early simulations did not precisely match what was observed in redshift surveys.

Cosmologists ran more simulations as numerical experiments: what happens if the simulation has a different amount of dark matter, or a different value of the Hubble constant? Is the pattern of the cosmic web quantitatively different? What happens if we include dark energy, or another form of dark matter like neutrinos? By comparing the patterns of the cosmic web seen in simulations versus in redshift surveys, the hope was that it would be possible to determine what the Universe was made of.

In the mid-1990s, the u.s. National Science Foundation's Grand Challenge Cosmology Consortium tasked a group of top simulators (including my advisor, Lars Hernquist) to understand the cosmic web and its observable properties using cosmological simulations. By considering all the available observations, including redshift surveys, this project had by the end homed in on a

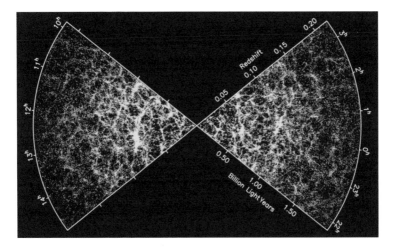

17 A map of the galaxy distribution around us from the 2dF Redshift Survey. We are at the centre, looking outwards 2 billion light years in two directions over regions of the sky covering more than five hours in right ascension. Each yellow dot represents a galaxy, with its distance measured from its redshift using the Hubble–Lemaître Law. Galaxies are not randomly spread across the sky, but instead are organized into large-scale structures up to 100 million light years across or more.

model with 20–40 per cent dark matter, a few per cent baryonic matter and the rest in dark energy as the most viable model. While the observations of Type Ia supernovae represented the final confirmation of dark energy, simulators had already reached a similar conclusion by comparing cosmological simulations to the cosmic web as seen in redshift surveys. The Type Ia supernovae provided a better measurement and more direct evidence for dark energy that didn't involve these newfangled numerical simulations, but it was reassuring to cosmologists that independent methods were all pointing towards ΛCDM as another example of concordance in the concordance cosmological model.

After the turn of the millennium, both redshift surveys and supercomputer simulations exploded in size and scope, as the cosmology community realized how valuable they could be. Key redshift surveys such as the Two Degree Field (2dF) Survey in Australia and the Sloan Digital Sky Survey (SDSS) in the United States mapped hundreds of times more galaxies than in those early redshift surveys. These redshift surveys, particularly SDSS, continue to this very day, providing forefront data that characterizes the cosmic web in remarkable detail. Today, the cosmic web is one of the main pillars of evidence supporting the modern concordance cosmology, providing measurements that are complementary to and independent of the CMB and Type Ia supernovae data.

WEAVING THE COSMIC WEB

Producing the cosmic web right out of the gate was a big win for cosmological simulations, and moved the wider community to take such simulations seriously. Particularly exciting was that simulations provided a complete filmic narrative of how it all came about, not just still-frames as seen from observations. In simulations, it is possible to trace back an individual object, or even an individual particle, and try to understand why it ended up where it did. As such, with simulations it became possible to answer the question that had intrigued cosmologists in the 1980s and '90s: how is the cosmic web woven?

The key new aspect that simulations elucidated was that the initial density perturbations seeded during the inflationary epoch are not spherical but rather ellipsoidal. Let's think about what happens to an ellipsoidally shaped density peak.

An ellipsoid has a short axis, a long axis and a middle axis. Being more dense than its surroundings, its gravity will attract matter around it. Zeldovich in the 1970s was the first to work out with mathematics how this would proceed. The shortest axis, which has the least distance to collapse and is closest to the peak of the matter and hence feels the most gravity, will collapse first. This will result in a two-dimensional structure known as a Zeldovich pancake, which corresponds to a sheet in the cosmic web.

Gravity doesn't stop, so the collapse will continue. The same scenario applies to the two remaining axes – the middle axis, being the shortest one left, will collapse next, leaving two axes compact while the third axis is still long; this resembles the shape of a cigar or a filament. Thus a filamentary structure is the second stage of collapse. In the final stage, the third axis collapses, and we get a node. This node will be ellipsoidal, retaining the initial information about the ellipsoidal nature of the density fluctuation.

The different large-scale structures in the cosmic web – the filaments, sheets and nodes – thus represent different stages of collapse due to gravitational instability (illus. 19). Regions that were the most over-dense have the strongest gravity, so they go through these stages quite quickly and rapidly collapse into nodes. Regions that are a bit less dense aren't all the way through the process of collapsing yet, and at any given time are still in the filamentary stage. Regions of very modest over-densities collapse quite slowly, since they don't contain as much gravity, so they haven't even made it past the Zeldovich pancake stage after 13.8 billion years.

Meanwhile, as all these over-dense regions gather up the surrounding matter, this leaves under-dense regions to form voids, giant volumes of cosmic space that have below-average cosmic density. As time passes, the rich over-dense regions become richer, while voids grow and become more prominent.

The formation of sheets, filaments and nodes is known as hierarchical structure formation. In hierarchical structure formation, the largest perturbations collapse first, and then grow by attracting and swallowing smaller objects around them. Thus coherent structures of matter in the Universe get larger over time, a process that continues even today.

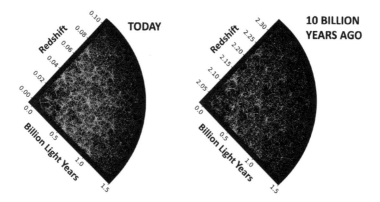

18 The two images show the distribution of galaxies in a cosmological simulation, viewed over a survey region with a geometry similar to (half) the 2dF Survey. The left image shows how such a survey would appear in this simulated universe at the present day. The cosmic web is evident, just like in the real galaxy distribution seen in 2dF. With a simulation, we can also look at the distribution as if we were living 10 billion years ago, as shown in the right image. The cosmic web is less evident at this earlier cosmic epoch, so although such large-scale structure was established early on, it has grown stronger over time as gravity pulls matter together.

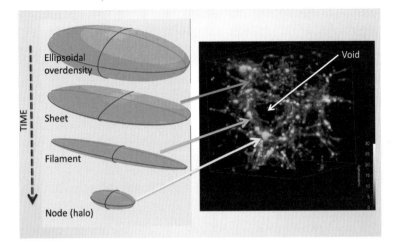

19 The formation of the cosmic web: an initial ellipsoidal over-density collapses along the short axis to form a sheet, then along the middle axis, forming a filament, and finally along the remaining axis into an ellipsoidal node hosting a dark matter halo. The sheets, filaments and nodes in the cosmic web represent different stages of ellipsoidal collapse at a given moment. The right panel illustrates these structures in the matter distribution from a cosmological simulation shown evolved to the present day. Filaments are prominently seen, sheets are diffuse so more difficult to see, and in between these structures are voids, regions where matter has been mostly evacuated. The nodes occur at the densest locations in the initial density field, while the voids form in the least dense regions.

Modern N-body simulations show the cosmic web in remarkable detail. Illustration 20 shows a snapshot of the Millenium-II N-body simulation, starting from the entire volume in the upper left, and then blowing up the inset region in each successive clockwise panel. The filaments and sheets are evident on large scales, and these can be seen feeding into the node in the middle where several filaments have converged. The patterns of this cosmic web emerging from the concordance cosmological model are in remarkably good agreement with modern redshift surveys, which provides an independent confirmation of the ΛCDM cosmological paradigm.

HALOS: THE SOLUTION

The final stage of hierarchical structure formation is when all three axes have collapsed into a node. But why does it *stop* collapsing? What is preventing the ellipsoidal node from continuing to pull itself into a tighter and denser configuration, until it finally forms a black hole? To understand this, let's take a trip to the Scottish National Museum.

Like many museums, the Scottish National Museum has one of those giant basins where when you roll a coin in from the edge, it slowly spirals down towards the centre, where it plummets down a hole. It's great fun for the kids, especially at the point when the coin gets close to the small hole in the centre and starts spinning around faster and faster in a frenzy of sound and fury, before finally disappearing into the circular rift.

The coin's movement and eventual drop are owing to friction. As it rolls on the surface, it makes noise and in other ways loses energy to its surroundings. These effects arise from electromagnetic forces – they come from the interactions between atoms of the coin touching the atoms of the basin and the air.

What does that have to do with the cosmic web? A dark matter particle in a node is a bit like that coin. It's trapped in the well, it isn't escaping, but it also isn't falling straight in. But dark matter has a property that distinguishes it from the coin: it doesn't suffer friction, because a dark matter particle does not interact electromagnetically. As such, it cannot fall down the hole, even though it feels gravity, because there is no friction to dissipate all that orbital energy. Imagine a sea of infinitesimally small dark matter particles in a giant frictionless marble basin, all zooming around, never hitting each other, never sinking to the centre, and you get the picture of what this looks like.

Such an object is known as a dark matter halo. A dark matter halo holds its ellipsoidal shape because the dark matter is prevented from falling in by its orbital energy. The orbital energy was obtained from the gravitational attraction by the mass in the cosmic web node. It's because of this lack of friction that dark matter does not sink down into black holes, but rather collects into large, fluffy halos.

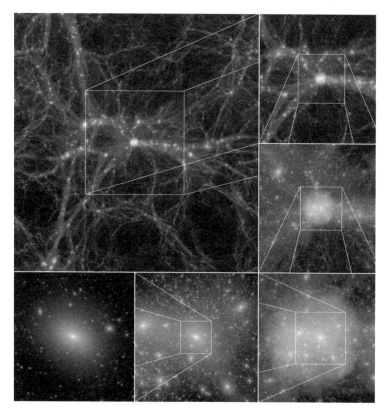

20 Cosmic web in the Millenium-II cosmological *N*-body simulation, zooming in from the largest scales through the cosmic web onto a single halo. The upper left box covers over a billion light years on each side, while the most zoomed-in region shown in the lower left covers around 10 million light years. Even in the most zoomed-in view, numerous halos of various sizes are seen, with the large central object showing the most massive halo in the region.

21   Galaxy cluster Abell 1689 contains thousands of galaxies concentrated in a large node of the
cosmic web. Its dark matter mass can be measured from galaxy velocities. It can also be mapped
using gravitational lensing of background galaxies distorted into arcs; this is shown in the blue
tint. The dark matter is in a more diffuse distribution, showing a dark matter halo around each
galaxy but also spread throughout the cluster.

The speeds we are talking about are not slow. In the halo of
the Milky Way, the typical speed of dark matter particles zooming around is 200 km/s. That's the distance from London to
Birmingham in a single second. This is the amount of velocity a
dark matter particle needs in order to hold itself up against the
massive amount of gravity generated by the Milky Way's entire
dark matter halo. Of course, individual speeds vary – as dark

matter approaches the centre of the halo, it speeds up, but like a coin that just misses the drain, it cannot stop at the bottom and fall in, but instead zooms past the centre and slows down as it moves outwards, continuing its tireless journey.

It's difficult to directly observe dark matter halos, since dark matter doesn't emit, reflect or absorb light. But there is a huge amount of evidence for their existence. The most direct way to 'see' halos is via gravitational lensing, where the light from background objects has its path bent by an intervening massive halo, like microlensing – which we discussed earlier (illus. 5) – only on a much larger scale. The more traditional approach is to use luminous tracers that are caught up in the halo's gravity, and likewise require large velocities to hold themselves up from falling in; this is the way that dark matter halos were first discovered, both by Zwicky in 1932, who measured the velocity of galaxies in the Coma Cluster, and by Rubin in the 1960s, who measured the velocities of gas and stars on the outskirts of galaxies.

Another way to measure the dark matter mass is to look at the temperature of the luminous gas in large halos. Unlike the dark matter, blobs of gas do not pass through each other but rather smash into each other, creating shocks that generate heat. The heat is enormous – the largest halos have temperatures up to 100 million Kelvin, purely from gravitationally driven shock heating. This gas can be detected via its bright X-ray emission.

The masses measured from these different ways for a given object typically agree to within 10–20 per cent. An example is shown in illustration 21, which depicts the galaxy cluster Abell 1689, a dark matter mass in excess of 1 million billion Suns. The total mass in the stars is about one-hundredth the mass in the dark matter, while the mass in the hot gas is around one-seventh as much. The remaining 85 per cent of the mass is then inferred to be in dark matter. This, and analogous data from other galaxy clusters, corroborates BAO measurements from the CMB confirming that five-sixths of the matter in the Universe is in the form of dark matter.

FROM HALOS TO GALAXIES

The Universe began as a homogeneous admixture of dark matter and baryonic (ordinary) matter, in roughly a 5:1 proportion by mass. Both components respond to gravity, and thus both are subject to gravitational instability. As such, the final collapsed halos contain not only dark matter, but baryonic matter. During the formation of the cosmic web, the baryonic matter was mostly along for the ride, dragged along by the gravity of the dominant dark component. Until a halo formed. Then things got interesting.

Let's go back to the coin basin, except now let's imagine that our orbiting object is a hydrogen atom. Unlike dark matter, a hydrogen atom has processes that allow it to dissipate energy. So, like the coin, it will eventually go down the drain. As a result, once baryonic matter gets inside the halo, it can separate from the dark matter and fall towards the centre, even while the dark matter is consigned to orbit around in a halo. This is the fundamental difference that explains why baryonic matter – that is, the elements in the periodic table – can form into galaxies and stars and planets and humans, while dark matter remains upheld in a diffuse halo. This is why we, and everything we see around us, are made of baryonic atoms and not dark matter.

The key physical process that separates baryonic matter from dark matter is known as radiative cooling, a process by which kinetic energy is converted into light energy (that is, radiation) through atomic physics. The light then streams out into the cosmos, carrying the energy away.

To understand how radiative cooling happens (illus. 22), you can harken back to the chemistry lesson in which you may have learned that a hydrogen atom consists of a proton orbited by an electron. Quantum mechanics tell us that the electron can only have discrete energies – for instance, an electron in a hydrogen atom can have an energy of –13.6 electron-volts (eV), or –3.4 eV, or –1.5 eV and so on, but it cannot have energies in between. Negative energies signify that the electron is bound – that is, it has too little energy to escape the pull of the proton owing to the opposite charge.

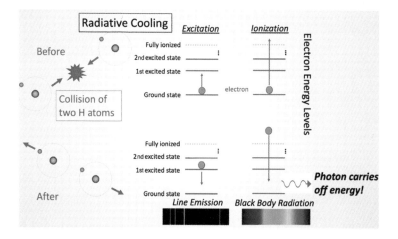

22 Collisions of hydrogen atoms lead to radiative cooling. The atoms either get excited (electrons moving up to a higher energy level) or ionized (electrons unbound from the proton altogether) due to the energy of the collision. Afterwards, the electron falls back down to a lower energy level and the energy is released as a photon (that is, a packet of light energy) – excitation results in emission lines, while ionization results in so-called black body radiation. The energy lost enables the hydrogen atoms to sink down into the centre of a halo. This is how radiative cooling converts the energy of motion into light that can stream off into space. We see this energy as light emitted from gas nebulae; in other words, this process is why objects in space shine.

Now, imagine two hydrogen atoms in the Milky Way's halo are zooming around at 200 km/s. Even though space is mostly empty, occasionally these atoms will smash into each other. Let's say their electrons are sitting in the ground state, which is the lowest possible energy level they can have. But in the collision, the electron gets kicked into a higher-energy state (called excitation), or maybe even entirely out of the atom (ionization).

Electrons, like English bulldog puppies, are lazy; they would rather live at the lowest-energy state possible. So after the collision happens, the excited electron heads back towards the ground state. To do so, it must lose energy. As Einstein surmised in 1905 (with the theory that ultimately won him the Nobel Prize), the way that electrons lose energy is by emitting photons – that is, packets of light. This light streams off into space, carrying with it some energy, lost from the atom forever. In this way, the kinetic energy of the collision gets transformed into radiative energy.

According to the law of conservation of energy, the atom must have lost some energy. This comes from the energy of its motion. So the atoms will slow down a bit. With repeated collisions, the atom loses more and more of its kinetic energy, and it can no longer hold itself up against the pull of gravity. Essentially, radiative cooling acts like friction for the coin in the basin, carrying away its energy and allowing it to fall towards the centre.

We refer to the average velocity of the gas as its temperature. So, effectively, the collision and subsequent emission of a photon cools the gas down to a lower temperature. That is why this process is called radiative cooling – 'radiative' because it emits electromagnetic radiation (light), and 'cooling' because the gas temperature drops. The above example for hydrogen is just one of many pathways for radiative cooling; other atoms, or even just free electrons, can also cool. In all cases, radiative cooling involves some kind of collision between two particles and subsequent radiation of a photon. This is a primary physical process that separates the baryonic matter from the dark matter.

Cooling is mostly unimportant in the diffuse cosmic web. But once gas gets into halos, it ramps up quickly. The reason is that collisions are much more frequent when the atoms are close together, that is, when the gas is dense. A collision requires that two atoms meet in a single place. If these atoms have some density n, then the probability that two particles will end up in exactly the same place is proportional to the density squared (n^2). Hence the rate of energy loss due to radiative cooling scales as the density of the gas squared. For instance, if there are three times as many hydrogen atoms in a given region of space, then the rate of energy loss via radiation will be $3^2 = 9$ times higher.

A simple way to envision this is a tic-tac-toe (noughts and crosses) board. Imagine you are allowed to roam horizontally along three columns (so you have a 'density' of one-third), and your friend roams vertically among three rows. You both randomly pick a box on the tic-tac-toe board at the same time. The chance that you will both pick the same box and have a 'collision' is (one-third)^2 = (one-ninth). This is why collisions have a probability that scales as the square of the density.

Clearly, collecting matter into dense regions is crucial for having large amounts of radiative cooling. Fortunately, gravitational instability does exactly this, collecting baryons (along with dark matter) into halos. It just so happens that the densities inside halos are high enough to get sufficient radiative cooling to allow the gas to lose significant energy, whereas the lower densities in the filaments and sheets cannot.

Because of radiative cooling, the baryonic matter inside a halo separates from the dark matter, spiralling down towards the centre of the halo to form a dense concentration of baryons surrounded by an extended halo of dark matter. Cosmologists refer to such an object as a galaxy.

A GALAXY IS BORN!

A cosmologist's definition of a galaxy is a collection of baryonic matter at the centre of a dark matter halo. But the same question arises here as in gravitational collapse: Why does the baryonic matter not continue cooling and losing energy ad infinitum, until it collapses into a black hole? What halts the collapse and causes the formation of a disc galaxy like the Milky Way?

Think of ice skaters. You've all seen this trick: an ice skater begins spinning with their arms outstretched, and as they pull their arms in, they start spinning faster and faster. This is a manifestation of the law of conservation of angular momentum. Since angular momentum is the product of the physical extent and the spin velocity, when the skater lowers their physical extent by drawing their arms in, the rate of spinning must increase in order to keep the same angular momentum.

What does this have to do with forming a galaxy? In a sense, the halo is like a slowly spinning ice skater whose arms are fully extended. As radiative cooling kicks in, the visible matter breaks away from the dark matter and starts falling towards the centre. As it does, its physical extent gets smaller – like a skater when they pull in their arms, the baryonic matter starts spinning faster and faster.

Now let's think about merry-go-rounds. You probably remember from childhood that it is difficult to move towards the centre of a spinning merry-go-round. This is because of centrifugal force – the idea that your body would prefer to move in a straight line

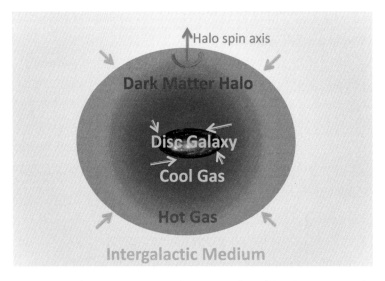

23 A disc galaxy forms within a dark matter halo. In the outskirts of the halo, the gas is heated by shocks as it falls in, creating a hot halo. Towards the centre, the increased density results in radiative cooling, allowing gas to cool and fall towards the centre. The halo is barely rotating, but as the gas falls in, it conserves angular momentum, spinning faster. This results in a flattened disc of gas, which is now dense enough to form stars and become a galaxy.

but you are forcing your body to turn by hanging on to the merry-go-round. You feel this as a force that flings you outwards, relative to the centre of the merry-go-round. In contrast, standing up versus sitting down on a merry-go-round is comparatively easy (assuming you aren't dizzy), because the centrifugal force only applies in the plane of the spin.

The same physics applies to galaxies. The baryonic matter falling towards the centre of the galaxy feels that same centrifugal 'merry-go-round' force. As it tries to fall in, it gets pushed out . . . but only in the plane of the spin. From the top and bottom, there is no centrifugal force, so in the vertical direction the visible matter is able to collapse into a thin sheet. Only in the rotational plane is this sheet held up from collapsing, which turns the collection of baryons into a flattened spinning structure. This is a disc galaxy (illus. 23).

The galaxy contains gas, and this gas is able to continue to cool and collect into molecular clouds. Within molecular clouds, stars are formed. This is itself a complex process and the subject of much study and debate, but suffice it to say, wherever there is a collection of cold dense gas in the Universe, we see stars forming. The stars then light up in the thin rotating disc, and voila, we've formed a disc galaxy (illus. 24).

IS GALAXY FORMATION SO SIMPLE?

We have connected all the dots leading us from the quantum fluctuations in the pre-inflationary Universe to the formation of a disc galaxy like the Milky Way. *N*-body simulations can predict where halos form and how much mass they have, and if we assume one-sixth is baryonic we can estimate how big the resulting disc galaxy in the centre will be. In this way, we can form an entire population of galaxies from a cosmological simulation.

This whole story of galaxy formation was worked out in the 1970s and '80s by some of the pioneers of galaxy formation theory, including Sir Martin Rees, Jeremiah Ostriker, Simon White and Carlos Frenk. When the first galaxy surveys came around, it was a golden opportunity for cosmologists to one-up the observers by showing that they had it all figured out already.

But there are more things in heaven and Earth than are dreamt of by cosmologists. This simple cosmologists' model for forming galaxies was a catastrophe. While it was able to qualitatively pro-duce some features of galaxies like rotating discs and even spiral arms, the galaxies did not resemble real galaxies in any detailed sense. Dispirited cosmologists soon realized that this cartoon

24 Examples of nearby disc galaxies, face-on and edge-on. On the left is M101, which is seen face-on from our vantage point. This shows a reddish central bulge and bluer spiral arms, with bright knots showing locations of new stars forming. On the right is NGC 4206, seen almost exactly edge-on, which illustrates how thin disc galaxies are. Our Milky Way is similar to these galaxies.

story for galaxies must be far from the full story of what makes galaxies look the way they do. There is more going on than just excited electrons and merry-go-rounds.

If the goal is to understand why the Universe looks the way it does, the problem of galaxy formation must be solved. It didn't take long for cosmologists to realize that forming realistic galaxies is going to involve a lot more physics, covering a wider range of scales, than anything encountered in astronomy before. Numerical simulations were the only viable way to ensure that so much physics could be accounted for within a single model. This leads us headlong into one of the most exciting and fastest-moving areas in astrophysics today: galaxy formation simulations.

Buckle up folks, it's about to get real!

4

A UNIVERSE OF GALAXIES

The night sky is full of stars. For millennia, humans have gazed at these stars every night as they move across the sky with comforting predictability and wondered where all of it came from. Ancient peoples invented stories of gods and heroes immortalized in the patterns of stars. They intertwined those stories with myths about the creation of the world and themselves. They developed complex astrological frameworks to assign those patterns to human characteristics, in what may be the most striking example of humans' instinctive need to connect to the cosmos. The stars in the night sky have been a source of wonder, awe and inspiration for humans since the dawn of our history.

Yet for all that, stars are just a fraction of all that we can see in the sky using today's telescopes. The vast majority of objects that we have catalogued in the sky are galaxies. These galaxies are not visible to the naked eye, except for a few which are very close by, such as the Large and Small Magellanic Clouds (visible from the Southern Hemisphere).

In the early twentieth century, a watershed moment in human history went largely unnoticed. It was then that increasingly large telescopes crossed the Rubicon, as images of the night sky began to show numerous faint fuzzy blobs of light in addition to bright point-like stars. These mysterious fuzzy nebulae, often having a whirlpool-like appearance, were seen in every direction in the sky, unlike stars that concentrated along the band of the Milky Way.

In 1920, two leading astronomers faced off in the first Great Debate in Astronomy at the Smithsonian Museum of Natural History. They argued the two leading views for the nature of these fuzzy nebulae. In one corner was the famous Harlow Shapley, who posited that these spiral nebulae were an unknown kind of gaseous object within our own Milky Way, with Andromeda being the nearest example. The primary argument in his favour was one of incredulity: if Andromeda were somehow outside the Milky Way, then assuming it rotated like a platter, its physical extent would be so large that the speed of stars in the outskirts would exceed the speed of light, and thus be in violation of Einstein's special relativity.

In the other corner was Heber Curtis, a less famous astronomer from Lick Observatory in the foothills above San Jose, California. Curtis argued that spiral nebulae were 'island universes' – as presciently envisioned by the philosopher Immanuel Kant in 1755 – entirely separate collections of stars lying well outside our own Milky Way. Their existence and numbers suggested that the Universe extended far beyond our own galaxy, making it larger than most astronomers had imagined up to that point.

The Great Debate wasn't conclusive, but it stimulated widespread interest in discovering the nature of these faint nebulae. The answer came just a few years later, from a familiar fellow: Edwin Hubble. To discover the expansion of the Universe, Hubble had to measure the distances to these nebulae. Hubble immediately realized that the distances placed these nebulae far beyond even the most optimistic estimates for the size of the Milky Way. This proved for certain that those nebulae were indeed distinct galaxies, collections of stars like Kant's island universes. It also proved that the Universe was immensely larger than our own Galaxy.

Today, the deepest images of the Universe that we can take come from the telescope that bears his name: the Hubble Space Telescope. Illustration 25 shows the Hubble eXtreme Deep Field (HXDF), one of the deepest pictures ever taken of our Universe. Within this image, which is smaller than the size of the nail on your little finger held at arm's length, you can easily pick out hundreds of objects by eye, and computers can pick out many thousands. Every single one of these objects is a galaxy, whose light is (predominantly) coming from the many billions of stars it contains. The galaxies in this image are seen at a wide range of cosmic epochs, from well-resolved nearby ones – galaxies seen as they are at the present time – to some of the earliest galaxies formed in the Universe, whose light has been travelling to us from a time less than a billion years after the Big Bang.

The HXDF illustrates why galaxies are the primary markers by which we map out the Universe and how it evolves over time: galaxies are the most clearly identifiable objects that we see in deep images of the cosmos. You can see from the HXDF that galaxies are as unique as snowflakes, coming in a wide variety of colours, shapes and sizes. This diversity is one of the wonders of the cosmos, and somehow must all arise from those tiny density fluctuations established during cosmic inflation and seen in the cosmic microwave background.

How does this all come about? How does such amazing diversity emerge from the one-part-in-a-million uniformity of the CMB? Understanding this process is tantamount to understanding why the Universe looks the way it does. This is the subject of arguably the largest and most active area of astronomy: galaxy formation and evolution. Let's learn a bit about galaxies, so we can understand what it is that we are trying to simulate.

THE HUBBLE SEQUENCE

Even in his first images, Edwin Hubble noticed that galaxies come in a variety of shapes and sizes. Not all galaxies appeared to be whirlpools like Harlow Shapley's spiral nebulae. Some galaxies were featureless circles, some were elongated ellipses and occasionally others looked like train wrecks. While for cosmologists, galaxies are nothing more than markers for the expansion rate of the Universe, Hubble was intrigued by the morphological diversity of the galaxy population. Faced with a plethora of new objects that he had never seen before, Hubble did the first thing that a good scientist does: he classified them into groups. This taxonomy of galaxies turned out to be remarkably useful, and to this day it bears his name, the Hubble sequence.

Hubble's classification scheme, presented in 1926 and sometimes called a 'tuning fork' diagram, separated galaxies into three main types: ellipticals, spirals and irregulars (see illus. 26). As these names suggest, the categories correspond to the visual appearance of these galaxies in the sky. Ellipticals are round-ish with varying elongation, spirals have whirlpool-like arms extending outwards from a central bulge, and irregulars are a catch-all for the few per cent of oddball galaxies that fall into neither category; typically these are either train-wreck galaxies undergoing a merger or dwarf galaxies undergoing a burst of star formation.

Within ellipticals, Hubble classified galaxies from E0 (perfectly round) to E7 (with an elongated shape). Within spirals, he classified galaxies by how many spiral arms they had and how tightly they were wound: *Sa* galaxies had a large bulge with many tightly wound arms; *Sb* and *Sc* had fewer and less-tight arms; and *Sd* had only two, loosely wound, arms and a small bulge or none at all. He also subdivided spirals into whether their central bulge

25 The Hubble eXtreme Deep Field (HXDF) is the deepest picture of the Universe ever assembled, constructed by combining images spanning ten years from the Hubble Space Telescope in a tiny patch of the sky less than 1 per cent of the area of the full Moon. Nearby galaxies are well resolved into round-ish red ellipticals or blue spiral discs, while distant galaxies are barely visible reddish pinpoints. More than 10,000 galaxies are detectable in this image, with the earliest one seen over 13.3 billion years ago. Such deep field images have revolutionized the exploration of galaxies across cosmic time.

was round or looked cigar-like, the latter called barred spirals (*SB*). In between spirals and ellipticals are so-called lenticular (or *S0*) galaxies, which have some hint of a disc but no spiral arms. The Milky Way is classified as an *Sbc* galaxy, that is, an unbarred spiral lying between the *Sb* and *Sc* classifications, though there is ongoing debate about whether it has a weak bar.

While Hubble only classified based on the morphology (or shape) of galaxies, this turned out to be prescient for many other galaxy properties. Observations over subsequent decades showed that spiral galaxies tend to be bluer in colour, have many young newly formed stars, contain lots of cooler gas, have modest central black holes and tend to live in the filaments of the cosmic web. In contrast, elliptical galaxies tend to be redder in colour, have no young stars, contain mostly hot (millions of Kelvin) gas, sport huge central black holes and tend to live in the nodes of the cosmic web.

The correlated properties along the Hubble sequence must arise from the way in which galaxies form and grow over time, so surely they must be telling us something fundamental about the physics of galaxy formation. Like the Greeks of old, modern galaxy formation theorists want to build a model that successfully explains all these properties and their attendant correlations. This monumental task stands at the forefront of modern astrophysics.

The great importance of understanding galaxies arises because galaxy formation lies at the nexus of numerous subfields of astronomy (illus. 27). Cosmology determines the patterns of the cosmic web within which galaxies form and grow. Galaxies themselves are made of stars, whose formation is itself an entire branch of astrophysics. Those stars produce the heavy elements that are necessary for planets like Earth and life as we know it to form. The nature of dark matter can influence galaxies as well. Most sizable galaxies contain a central supermassive black hole whose origin and growth remain hotly debated. We will later discuss a bevy of so-called feedback processes, whereby energetic release from dying stars and black holes can impact how galaxies evolve. The interconnectedness of astronomy from the smallest cosmic scales to the largest is one of its most challenging aspects, and nowhere is this more apparent than when studying the formation and evolution of galaxies.

The goal of galaxy formation theory is to develop a physics-based model, starting from the Big Bang, that not only yields the correct number of galaxies in each class, but correlations between galaxy morphology and their ongoing star formation activity, their environment within the cosmic web, the properties of their gas and their central supermassive black holes, and anything else we can observe about them with our bevy of modern telescopes. We want our model to successfully predict not only the galaxies as they exist today, but the galaxies that populated the earliest epoch of the Universe, and all the galaxies in the time between (see illus. 28).

Given the complexities involved, it probably comes as no surprise that galaxy formation theory relies heavily on numerical simulations. But if we want to model galaxies directly, we cannot

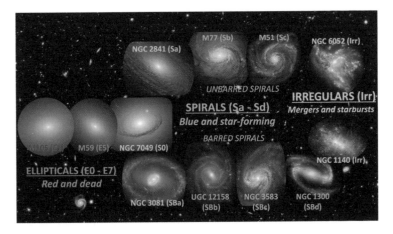

26   The Hubble sequence, showing elliptical, spiral (unbarred and barred) and irregular galaxies, with example galaxies shown for each Hubble type. The classification scheme separates galaxies by morphology, but it happens that this correlates with a number of other galaxy properties, including how rapidly a galaxy is currently forming new stars. This makes the Hubble sequence a highly informative framework for investigating galaxy formation and evolution. The background shows the Hubble Deep Field, the first deep field taken with the Hubble Space Telescope in a visionary project by then-director Robert Williams that launched the deep field revolution.

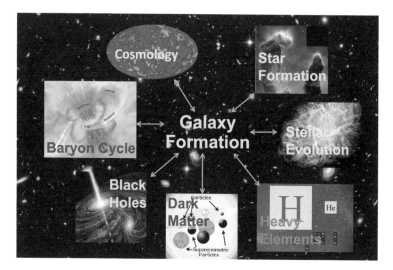

27 Galaxy formation stands at the crossroads of many branches of astrophysics, from cosmology on large scales to star formation and black hole growth on small scales.

simply model the dark matter alone. Galaxies are made of baryons, which undergo processes that dark matter does not, such as radiative cooling and supersonic shocks. If we want to model all these effects to form galaxies in a simulation, we are going to have to put a lot more physics into our computers than just gravity. As we will see, this makes things much harder.

GASTROPHYSICS

The problem of simulating galaxies is in a sense opposite to the problem of simulating dark matter: for dark matter, we have no idea what it is, but we know how it behaves since (insofar as we know) it responds only to gravity. In contrast, the visible parts of galaxies are made up of ordinary baryonic matter such as hydrogen and helium, so we know exactly what it is made of, but it can behave in extremely complex ways.

The starting point for simulating galaxies is the dark matter. We know the Universe contains it and we know how to model it. We can place a whole bunch of dark matter particles into a simulation box and run it forward in time until today using the force of gravity. While there are a lot of technical complications, as we described earlier, this is now mostly a solved problem.

A galaxy formation simulation operates on the same principles. But now, instead of just dark matter, we must include a second type of particle in our simulation box: a gas particle, representing an admixture of three-quarters hydrogen and one-quarter helium arising from Big Bang nucleosynthesis. Since gas is a fluid, galaxy formation simulations are often referred to as hydrodynamic simulations. Embedding such gas particles within a cosmological *N*-body simulation then makes it a cosmological hydrodynamic simulation. This is the numerical technique that is most often used to model the formation and evolution of galaxies today.

A gas particle responds not only to gravity, but to various other physical processes, such as gas pressure, radiative cooling, friction, shock heating, magnetic fields, nuclear fusion, chemistry, photo-ionization and more. As you can see, once we include gas into a cosmological simulation, the halcyon days of One Force to Rule Them All are long gone – there is now all this complicated 'gastrophysics' that we must include. Sure, it only affects one-sixth of the cosmic mass, but it happens to be the one-sixth that we can directly see!

Not only that, galaxies also have stars and often a central supermassive black hole. How do we know when our simulated galaxy should form a star or a black hole? This is tricky, since the formation of stars and black holes is itself not well understood. But if we want to produce realistic-looking galaxies containing all these components, we must somehow include their formation in our simulation as well.

Yet for all this, progress over the last two decades in galaxy formation simulations has been nothing short of spectacular. When I first began in this field prior to the turn of the millennium, galaxy formation simulations couldn't reproduce even the most basic galaxy demographics close to correctly. Within two decades, galaxy formation simulations are now routinely producing stunningly realistic representations of essentially the entire galaxy population over cosmic time, and leading the way towards understanding how the Universe came to look the way it does. Let's see how this remarkable journey unfolded.

THE HYDRODYNAMICS WARS

When intrepid galaxy formation simulators in the late 1980s began to think about how to simulate a galaxy, the basic idea would be to 'simply' add in the gastrophysics to N-body simulation. The first step towards this is to model the gas pressure forces on a computer. This already caused quite a kerfuffle.

One class of algorithms, known as Eulerian codes, partitioned space into a fixed Cartesian grid analogous to a particle-mesh gravity code, and the gas pressure forces were computed via the pressure across each adjoining pair of cell faces. However, like with a PM code, the resolution of the simulation was limited by the grid cell size, so it was not ideal for following gas down into a dense concentration at the centre where a galaxy would form. To mitigate this, adaptive mesh refinement (AMR) was invented. In AMR, cells that collect a lot of gas were further refined into smaller grids, creating grids-within-grids. This was a bookkeeping nightmare, but eventually several research groups made this

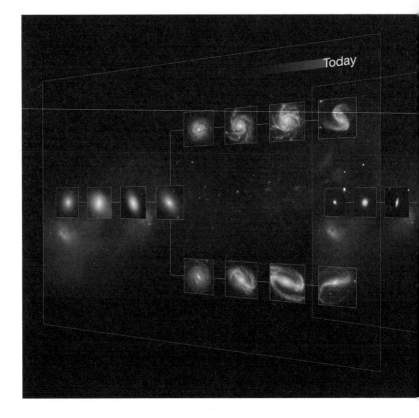

work, and this enabled increased resolution in the denser regions and retained spatial adaptivity.

A more annoying issue with AMR is that, when trying to move gas through a grid in an arbitrary direction, the grid generated artefacts due to the rigid Cartesian geometry. This was not ideal for modelling a rotating disc, whose circular orbits did not match well onto a cubical mesh. Moreover, such AMR codes tended to be quite slow, owing to the bookkeeping overhead of dealing with sub-meshes.

In the 1970s, a technique called smoothed particle hydrodynamics (SPH) was developed. In SPH, there was no grid at all; individual particles can be regarded as fuzzy blobs of gas, whose

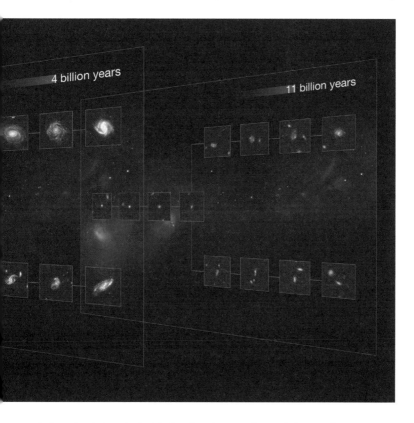

28 Example galaxies going back in time. From the CANDELS Survey, the largest galaxy survey ever done with the Hubble Space Telescope, the galaxies are arranged into a Hubble sequence at 0, 4 and 11 billion years ago. Going back in time, galaxies are smaller and more irregular, and look redder in general because their light has been redshifted due to cosmic expansion.

properties such as density and temperature are defined by averaging over neighbouring particles in a sphere of a radius given by the smoothing length. This is set to encompass a fixed number of particles (usually 50–100), so that in dense regions with lots of particles the smoothing length is small, giving high resolution, while in low-density regions the smoothing length is large and the resolution is poorer. This makes SPH naturally adaptive, like a tree code.

While attractive, SPH had its own set of issues. Crucially, SPH cannot represent discontinuities such as shocks very well, because the smoothing tends to wash out sharp features. Mesh codes, with sharp boundaries between grid cells, handle shocks and discontinuities better. SPH also had difficulty evolving a rotating disc where the gas had different rotation rates at different radii, because the smoothing intermingled gas at different radii to cause unwanted friction.

And so began the hydrodynamics wars. The SPH adherents claimed that the drawbacks of not handling discontinuities well were not so important for galaxy formation, while SPH's advantages of adaptivity and lack of grid artefacts made it superior. The mesh hydrodynamics crowd countered that if one wants to model gas flows properly, handling shocks accurately is crucial, and thus AMR codes are preferred despite being unwieldy and computationally expensive.

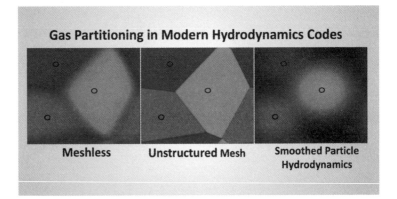

Gas Partitioning in Modern Hydrodynamics Codes

| Meshless | Unstructured Mesh | Smoothed Particle Hydrodynamics |

29 Different hydrodynamics schemes discretize the smooth gas field in different ways. Three commonly used schemes are illustrated here. The right panel shows SPH, where gas particles are envisioned as spherical blobs with overlapping fuzzy outskirts. The middle panel shows an unstructured mesh code, where cells are 'assigned' the space that is closest to their centres. The left panel shows a 'meshless' ALE scheme, which uses a particle-like representation of an unstructured mesh. Each scheme has different strengths and weaknesses.

To date, neither side has emerged victorious in the hydro-dynamics wars; both AMR and SPH codes remain widely used. A promising development in the last few years is that new codes have emerged which aim to marry the best of both worlds. Un-structured mesh codes use a mesh, except the mesh deforms over time to follow the mass, thereby allowing it to be adaptive like SPH but offering the shock-capturing abilities of an Eulerian code. Another class is the so-called arbitrary Lagrangian/Eulerian (ALE) codes, which employ a particle representation but nonetheless use shock-capturing schemes found in Eulerian codes, such as so-called meshless hydrodynamics schemes (illus. 29). Improving computational gas dynamics is an active area of research, as the hydrodynamics wars rage on to this day.

THE DISC GALAXY CHALLENGE

Armed with their shiny new hydrodynamics codes, in the 1990s galaxy formation simulators turned their attention to answering the first question that comes to mind: how does a disc galaxy like our own Milky Way form?

In the scenario outlined in Chapter Three, the basic picture is supposed to go as follows: from a slowly rotating dark matter halo containing an admixture of dark matter and gas, the gas radi-atively cools and falls towards the centre, and angular momentum conservation results in the gas settling into a flattened disc. The physics here is nothing but gravity, gas pressure and radiative cooling – all pre-twentieth-century physics. This should be simple enough, so surely it can be represented on a computer.

So confident were these simulators that they even went out on a limb to include some new physics in order to form a more realistic galaxy: star formation. In real galaxies there is a tight correlation seen between the gas surface density and the rate of star formation; this is known as the Kennicutt–Schmidt relation. Since simulations predict the surface density of gas, the Kennicutt–Schmidt relation can then be used to empirically convert gas into stars using this star formation rate, thereby mimicking what the real Universe does (despite a lack of understanding of the details). By implementing the Kennicutt–Schmidt relation into their hydrodynamics code, simulators could form galaxies that included dark matter, gas and now stars as well – the main ingredients seen in real galaxies.

The very first simulations dispensed with the full cosmological context for simplicity, and instead began with an isolated rotating region containing an admixture of gas and dark matter. What was seen largely followed expectations. The gas collapsed towards the middle of the halo, spun up due to angular momentum conservation, and formed into a disc, which formed stars. Success! At least superficially, it seemed that computers could produce a Milky Way-like disc galaxy.

But the cosmology community was not impressed. This was because these isolated galaxy simulations had not accounted for hierarchical structure formation. Halos are constantly accreting dark matter and gas from their surroundings via gravitational instability, and occasionally even merging with other halos. While the first isolated galaxy simulations produced realistic-looking disc galaxies, they did so under conditions that would never actually occur in the real Universe.

To do this properly, hydrodynamics would need to be implemented into full cosmological simulations. The gravity-only N-body simulations were already computationally expensive, but including hydrodynamics required evolving a new set of particles with many more gastrophysics processes. Even worse, if one wanted to resolve the fine details of a galaxy disc, one had to make the softening length quite small, which means the timesteps became small. Thus evolving over the full 13-plus billion years now took millions of timesteps instead of thousands, proportionally increasing the computational cost. But with Moore's Law in full swing, it was just a matter of time before the computers became powerful enough to handle this. (This law is the conjecture from semiconductor pioneer Gordon Moore in 1965 that the number of transitors in a computer chip – and hence computational speed – doubles every two years; it has more or less held true until recently.)

In the 1990s, cosmological hydrodynamic simulations of galaxy formation burst onto the scene. The great hope was that, rather than just producing dark matter halos in the cosmic web, computers would now produce full-fledged galaxies dotting the filaments, sheets and nodes, complete with stars and gas in various phases. The halos would form disc galaxies within them, and those discs would merge in collisions that would disrupt the discs and turn them into ellipticals. With this, the entire galaxy population in all its Hubble sequence glory would magically arise from the tiny matter fluctuations in the early Universe, and everything would be solved. These were heady times in which high hopes abounded.

But as so often happens, nature did not cooperate with these best-laid plans. Regardless of hydrodynamic scheme, the galaxy populations formed in these early cosmological hydrodynamic simulations bore, rather disappointingly, little resemblance to real galaxies.

TOO BIG, TOO FAST, TOO MANY

In cosmology, we have seen that halos start out small and grow hierarchically by drawing in and swallowing other, smaller halos from their surroundings. If these newly merged halos contain galaxies, then their galaxies will also eventually sink to the centre and merge. During simulations of the early Universe, when small halos abound and the Universe is more compact, such merging is extremely frequent.

This rapid merging is disastrous to discs. Delicate spiral discs cannot form and be sustained if they are constantly being bombarded with giant lumps of gas and stars. Bombarded galaxies end up being more like irregulars or ellipticals, with at most only small nascent discs around them.

As time passes, the universe expands, galaxies move further away from each other, and the merger rate drops. In the last half of cosmic time, galaxies can indeed follow more closely the simple isolated halo scenario. The trouble is, over the first half of cosmic time, they've already formed large central collections of stars in their bulge.

Due to this early bombardment, the first simulations of galaxy formation all ended up with overly large bulges. On the Hubble sequence, it was difficult to produce anything beyond a galaxy with Hubble type *Sa*; no *Sc*'s or *Sd*'s, hardly even any *Sbc*'s like the Milky Way. In contrast, in the real Universe, galaxies like the Milky Way are among the most common types of spiral galaxy. A galaxy formation model doesn't inspire a lot of confidence if it cannot even produce the very galaxy we live in.

But it gets worse. Remember Vera Rubin's measurements of flat galaxy rotation curves which indicated large amounts of dark matter? Simulated galaxies did not look like this. Instead, because of their large central bulge, the rotation curves started to look a lot more Keplerian, that is, like the Solar System's: they rose sharply in the inner parts where the overgrown bulge was, and then dropped off in the outer parts. It wasn't that the simulations didn't contain dark matter; they did. The trouble was the baryons, which ended up far too centrally concentrated as compared to real galaxies. Moreover, the rotation curve also peaked at a velocity far too high compared to real galaxies. For instance, a simulated Milky Way-sized galaxy had a peak rotation speed of more than 300 km/s, instead of the measured value of 220 km/s. The overly large central bulge, caused by lots of early merging, also messed up galaxy rotation curves.

But that's not all. Forget about morphology or rotation curves or anything fancy, just the sheer number of stars in galaxies was far too high. This was seen using a technique called abundance matching, which takes the halos in an N-body simulation and populates them with observed galaxies that have the same abundance (or number density). One can then measure the mass of stars in those observed galaxies to get the typical mass of stars within halos in the real Universe. This can be compared to how many stars are in a halo in the simulation. Early galaxy formation simulations overproduced the number of stars in halos – and not by a small amount either, sometimes by up to a factor of ten or more. There was no way the real Universe contained as many stars as predicted in these early simulations. Even more disastrously, as simulations improved in resolution, higher densities could be resolved and the radiative cooling rates became faster, so the problem only worsened.

This overproduction of stars was so ubiquitous and troubling that it was given a name: the overcooling problem. Pen-and-paper models of galaxy evolution in the 1970s had already hinted at overcooling, but by the early 2000s, state-of-the-art galaxy formation simulations had demonstrated that it was a serious problem that wasn't going away. Clearly, there was some basic and crucial ingredient that was missing in galaxy formation models. What was it?

PUT IT IN THE FEEDBACK BOX

The overcooling problem was simple to state: early galaxies grew too many stars, spun too fast and became too bulge-dominated. Solving it, on the other hand, was not nearly so easy. The physics leading to overcooling was bone simple, just gravity and radiative cooling; it seemed unlikely (though not impossible) that either of those could be wrong. If we wanted to prevent overcooling, we had to include some new physical process(es) that counteracted gravity, for example, or suppressed radiative cooling. What could we be missing?

To reduce early star formation in our simulated universe we need to do one (or both) of two things: stop the gas from cooling onto galaxies, or eject the gas before it can form into stars. In either case, it comes down to a question of energy. If you want to throw a ball upwards, you need energy. If you want to prevent your hot food from cooling down, you need energy. The same principles apply to galaxies. The energy needs to come from somewhere. So let's play detective and look around for possible sources of energy.

An obvious place to look for energy output is stars. Galaxies are full of them. Stars in the sky are like stars in Hollywood: the big ones shine bright, live fast and die hard. The dying part is the most interesting, because if a star is more massive than about eight times the mass of the Sun, it doesn't go gently into that

6 Nov 2014 | 25 March 2015 | 12 Nov 2015 | 8 April 2016 | 12 October 2016

30 The Hubble Space Telescope captured a supernova exploding in nearby starburst galaxy Messier 82, and tracked its light echo over several years. The cumulative energy from millions of such supernovae going off over millions of years drives the galactic outflows seen as the reddish material streaming outwards along the poles. This expulsion removes gas from the reservoir available to form new stars, and thus self-regulates the growth of the galaxy.

good night. Instead, these rare massive stars die in a spectacular explosion known as a supernova.

Each star that goes supernova releases about 10^{51} ergs of energy. To put that in perspective, a single supernova in a few minutes releases millions of times more energy than our Sun produces over its entire 10-billion-year lifetime. Supernovae are so bright that they can outshine an entire galaxy for up to a few months. Even though less than 1 per cent of all stars are big enough to go supernova, given the billions and billions of stars in a galaxy, even a small fraction of them dying as supernovae provides a huge amount of energy (illus. 30).

This energy results in supernova feedback. The word 'feedback' in galaxy formation refers to the energy released by a process associated with the growth of a galaxy, which serves to retard the galaxy's growth, sort of like self-regulation. In the 1980s, Avishai Dekel and Joe Silk estimated that supernova feedback was so energetic that galaxies smaller than around one-tenth of the size of the Milky Way could have all the gas in their entire galaxy blown out. For larger galaxies, such as the Milky Way, however, supernova feedback was less effective because the stronger gravity was able to prevent the gas from escaping.

When overcooling was recognized as a major problem in galaxy formation, the first thought was that surely supernova feedback must be the key to solving it. While it would only work on smaller galaxies that form stars very rapidly, thereby creating lots of supernovae, that's okay, because small, rapidly growing galaxies at early cosmic epochs are exactly where the overcooling problem starts. Thus even if Dekel and Silk's model didn't affect the Milky Way now, it would have been effective back in the early Universe when the nascent Milky Way was much smaller, effectively nipping overcooling in the bud. Perhaps the answer was blowing in the supernova-driven wind.

In the late 1990s, armed with spiffy new hydrodynamics codes, Mordecai-Mark Mac Low and Andrea Ferrara decided to try directly simulating the process of supernova feedback expelling gas, hoping to explain overcooling and solve galaxy formation. However, in what is surely becoming a familiar refrain, it did not go as planned.

COMPUTERS GOING SUPERNOVA

You've probably heard of spherical cow jokes. The joke is that a farmer wants to know how much he can charge for his milk, and the theoretical physicist begins by saying, 'Assume a spherical cow.' The joke is that physicists often begin by making simplifying assumptions that are so far divorced from reality, they have no practical value whatsoever. Okay, not super funny, but physicists have an odd sense of humour.

Dekel and Silk had assumed a spherical cow – or in their case, spherical supernova feedback. They had assumed that exploding supernovae push the gas in the galaxy outwards in a perfectly spherical manner. This is a reasonable approach, and sometimes the spherical cow is quite insightful. In this case, however, it turned out to be a bit too divorced from reality.

Anyone who has seen a Hollywood film knows that an explosion doesn't explode spherically if there is anything confining it. Instead, the exploding material follows the path of least resistance – usually down an alley or a tunnel, where the hero barely outruns the intense fireball before diving into a ditch that the fireball miraculously passes right over. As a rule, Hollywood films are not the best instructional videos for someone studying physics, but in terms of explosions following the path of least resistance, they win the physics derby.

As Mac Low, Ferrara and others discovered, three-dimensional simulations of supernova feedback within galaxies showed exactly this scenario: the supernovae accelerated gas to extremely high velocities, hundreds of kilometres per second, but the explosion followed the path of least resistance out of the galaxy, carrying very little mass. All the heavy and dense gas that leads to overcooling mostly stayed in place. At least in this simple scenario, supernova feedback hardly made a dent in overcooling. Dekel and Silk had found an answer to where the energy comes from, which is step one to solving overcooling. But they had not solved step two, which is how to deposit all that energy in a way that blows out the gas. There must be another piece to this puzzle.

Fortunately, supernovae are not the only source of energy coming from stars. Massive stars during their life put out a huge amount of energy in high-energy radiation, which can be absorbed by surrounding gas, heating it and generating radiation pressure which pushes the gas outwards. Encouragingly, radiation from stars is spherical, so radiation pressure can provide exactly the sort of spherical cow we need to clear out gas from a galaxy and suppress star formation. The downside is that it only provides a slow and gentle push, not nearly enough to achieve the velocities needed to escape from a galaxy's gravitational pull. There are other energy sources around as well, such as stellar winds and cosmic rays (that is, high-energy particles emitted by stars like the Sun that on Earth cause beautiful aurorae), that also tend to push outwards more spherically. Likely, the final answer is a combination of all these energy sources acting in concert, but exactly how that symphony is played remains a mystery.

Recent simulations redoing the Mac Low and Ferrara experiment have greatly improved in sophistication, such as those from the SILCC project (see illus. 31). They find that cosmic rays can be important for lifting gas out of galaxies, in conjunction with supernovae explosions. While an improvement, even these state-of-the-art simulations do not achieve the outflow rates that are needed to solve overcooling. The transition from modelling stellar feedback as a spherical cow to a real cow remains a work in progress in galaxy formation and evolution.

ZEN AND THE ART OF SUBGRID PHYSICS

The situation seems a bit hopeless. If processes associated with individual stars such as supernovae are critical to forming realistic galaxies, how can we ever manage to model this within a cosmological volume? As discussed earlier, a cosmological sim-ulation does not have nearly enough dynamic range to represent individual stars while maintaining a large enough volume to reliably represent the cosmic web. Worse yet, even if we could achieve that dynamic range, we don't know how to generate enough outflows to solve overcooling. How can we include stel-lar feedback in our cosmological hydrodynamic simulations to produce a realistic galaxy population if we don't even understand how feedback works?

One approach used in physics is to describe the global effects of complicated processes using an effective theory. An effective theory is a description of the overall macroscopic effects of numerous microscopic processes. A familiar example is weather – individual clouds have many unique characteristics, but gen-erally it is still possible to study them to predict when it's going to rain. Another more physics-oriented example is the theory of thermodynamics, the study of heat and pressure. Thermody-namics provides a model that effectively describes how heat is transported through substances and how it causes pressure, with-out directly dealing with the underlying microscopic processes of individual atoms bouncing around. This allows us to build car engines and refrigerators by effectively representing the impact of small-scale processes on the larger scales that we care about. Effective theories thus allow physicists to focus on the scales of interest, while still accounting for small-scale effects in a statistical sense.

This is exactly the situation we find ourselves in – we would like to model galaxies on large cosmological scales, while account-ing for the global effects of small-scale feedback processes such as supernova feedback. In galaxy formation simulations, the effective theory for such small-scale processes is called subgrid

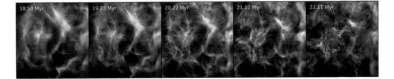

31 State-of-the-art simulation of supernovae exploding within a molecular cloud deep inside a galaxy. Over several millions of years (Myr), the supernovae generate blast waves that heat the surrounding gas. The cumulative pressure eventually generates an outflow with sufficient energy to escape from the galaxy, as seen in M82 (illus. 30). This simulation, created by the SILCC team, includes a wide range of up-to-date physical processes, including magnetic fields, cosmic rays and radiation pressure, but still has difficulty driving enough outflowing material to solve the overcooling problem.

physics. Unfortunately, this effective theory is not nearly as well developed as something like thermodynamics, since even the small-scale processes are extraordinarily complex and not fully understood. It is when including subgrid physics that galaxy formation simulations veer from rigorous science towards speculative philosophy.

There are three broad philosophies when including subgrid physics in numerical simulations:

The reality check approach: If we can observe the relevant process happening in the real Universe, we can try to mimic that behaviour in our simulation, even if we don't exactly know why it happens.

The multi-scale approach: We can use high-resolution simulations to depict how feedback works on small scales, and encapsulate its large-scale effects in simple equations that we implement into our cosmological simulation.

> *The 'just tune, baby' approach:* We can implement an
> extremely simplistic model for the effects of feedback
> with some free parameters, and vary those parameters
> by trial and error until we get a galaxy population that
> resembles what we observe.

Clearly, none of these approaches are fully satisfactory; none
of them directly model the physics of feedback, only the coarse
large-scale effects of feedback. But maybe that's enough to get
us past the small-scale physics problem and get on with studying
cosmological galaxy formation.

The 'reality check' approach is what was used for modelling
star formation. The Kennicutt–Schmidt Law is based on observations of a correlation between gas density and star formation
rate, which can be implemented straightforwardly into simulations to determine how much gas forms into stars. Unfortunately,
in the case of stellar feedback, observations of the amount and
velocity of galactic winds are highly uncertain, so no equivalently
robust relation is as yet known.

From a theoretical perspective, the multi-scale approach
seems the most gratifying. In that case, one is using physics-based
simulations at all scales, and therefore it is still a fully theoretical
prediction of a particular model. The downside of this is that the
high-resolution simulations themselves are extremely challenging and often inconclusive, such as supernova feedback simulations
that still have difficulty generating the (apparently) required level
of outflows. Moreover, it can be tricky to stitch together information from simulations at different resolutions. And if the final
simulation doesn't produce realistic galaxies, then there is ambi-
guity in terms of which model was wrong – is the cosmological
model at fault, or is it the high-resolution model which is feeding

information into the cosmological model that is flawed? This can make it difficult to learn how to improve from the failures and discrepancies. Nonetheless many simulators pursue this approach simply for its (relative) purity.

Other simulators prefer the 'just tune, baby' approach, in large part because it is agnostic. If feedback is a black box that prevents star formation, perhaps it is best to use the KISS (keep it simple, stupid) approach and parameterize our ignorance with simple formulae, then tune the parameters to get the answer we want, such as for solving overcooling.[46] The downside is that one doesn't necessarily learn much about the physics of galaxy evolution, one only learns about the parameters within the particular chosen framework, which might be overly simplistic and thus physically meaningless. Moreover, if we have already found the answer we want by tuning our free parameters, then how can we claim to have a predictive model?

These are the dilemmas that a cosmological simulator must face in order to design a galaxy formation simulation. Without a rigorous way forward, there is much debate, even to this day, as to which approach is the best way. In the end, what has happened is that different groups emphasize different approaches for different aspects of feedback.

This is the main difference between the perhaps dozen or more research groups around the world currently running galaxy formation simulation campaigns – they all use different approaches to model subgrid physics processes such as supernova feedback. The hope is that the different approaches, despite being tuned to match some observations, will yield divergent predictions for other data sets, which can be used to rule out models and home in on the most realistic set of physical processes describing galaxy formation and evolution. But that day seems as yet far off.

GALACTIC WINDS COME TO COSMOLOGICAL SIMULATIONS

The first simulations to include galactic winds in the early 2000s went with the 'just tune, baby' approach, for lack of better guidance. Yet these first results already demonstrated the huge impact that galactic winds can have on galaxies.

Volker Springel and Lars Hernquist were the first to include ubiquitous galactic winds (putatively driven by supernovae) in cosmological galaxy formation simulations. This was done in a very simple way: they kicked particles out of galaxies by imparting some specified velocity in a random direction, at a rate that is a set multiple of the star formation rate. The value of this multiple came to be known as the mass loading factor, often denoted by the Greek letter η, which is defined as the ratio of the mass outflow rate to the star formation rate. In their approach, for every gas particle that forms into a star, η gas particles are ejected out of the galaxy.

The mass loading factor has emerged as the single most important parameter governing how outflows self-regulate galaxy growth. By literally kicking cold gas out of galaxies, it reduces the fuel available to form stars, suppressing early galaxy formation and solving overcooling.

Springel and Hernquist tuned η such that their simulation roughly matched the observed total amount of stellar mass formed in the Universe. The value they arrived at was $\eta = 2$. This was already quite surprising – in order to solve overcooling, Springel and Hernquist's model required that for every star ever formed in the entire Universe, twice as much mass has been flung back out into intergalactic space. Think about all the mass in all the stars in a galaxy, and now imagine that twice that amount of mass in gas has been ejected from the galaxy back into intergalactic space. That is a lot of gas being strewn about the universe.

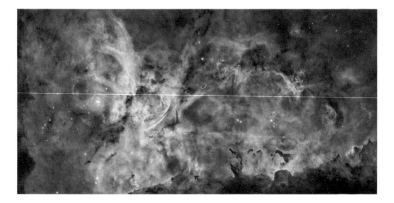

32   A Hubble Space Telescope mosaic of the Carina Nebula, a region of active star formation in the Milky Way approximately 7,500 light years from Earth. The hot young stars along with recently exploded supernovae dump enormous amounts of energy into the surrounding gas, lighting it up in a beautiful mess of colour and motion. In a galaxy formation simulation, this entire image would fit within a single resolution element. There is no practical way to capture all this complexity in a cosmological simulation, even though the effects of supernovae and radiation are felt on cosmological scales. The only practical way to include the effects of such processes is via subgrid physics.

The idea of massive amounts of gas exploding out of galaxies flies in the face of the fact that, when you look at an image of a galaxy, it appears to be a serene and delicate spiral. How can it contain explosions that are so strong as to accelerate such huge amounts of material to many hundreds of kilometres per second? Wouldn't such violence destroy spiral structure? Wouldn't we observe some impact on the surrounding halo? Wouldn't we directly see such large amounts of stuff spewing out of galaxies?

As it happens, observations of galactic winds were progressing apace to provide some guidance. Until the mid-2000s, galactic winds had only been observed in the most extreme starbursts, galaxies undergoing short-lived episodes of vigorous star formation leading to enormous numbers of supernovae exploding (like M82; see illus. 30). Such starbursts are very rare, and are typically triggered by the merger of two smaller galaxies. For most normal galaxies, there was no evidence of any material being ejected. So at the time it was proposed, Springel and Hernquist's idea of solving overcooling by having every star-forming galaxy eject double its mass in stars at high velocities seemed wildly discordant with observations of real galaxies.

But that would soon change. As galaxy formation theorists began harping on the importance of galactic winds, observational astronomers came up with clever ways to look for them. One particularly successful technique was to take a high-resolution spectrum of a galaxy, and look for gas that is blueshifted relative to the galaxy. The idea is that we would see ejected gas flowing

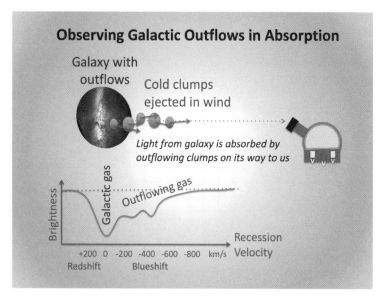

33 Seeing outflows via blueshifted gas. A spectrum of a galaxy shows absorption due to cold gas located in front of the shining stars. This has two components: (i) gas inside the galaxy, which is limited to roughly ±200 km/s owing to gas rotating in the galaxy's disc; and (ii) gas moving in an outflow along the line of sight, which can have much higher velocities. The outflow speed can be measured as the maximum speed where absorption from outflowing gas can be noticed; in the example shown this is about 800 km/s, which is typical of early star-forming galaxies.

towards us, out of the galaxy. By measuring the blueshift relative to its galaxy, it would be possible to determine the velocity of the outflow, as schematically depicted in illustration 33.

At early cosmic times, the Universe was smaller and denser, which means radiative cooling was faster, and thus star formation rates were higher, leading to more supernovae. One expects then that stellar winds might be more common. Sure enough, Chuck Steidel, Max Pettini and colleagues used the 10-metre Keck telescope atop Mauna Kea to detect outflows in distant early galaxies using blueshifted absorption lines. What was particularly remarkable was that they saw such outflows in virtually every early galaxy they looked at. At least at early cosmic epochs, outflows were not some rare phenomenon happening only in oddball galaxies; they were the rule rather than the exception.

The speeds of these outflows were typically between several hundred and a thousand kilometres per second. Blueshifted gas moving at these speeds was surely outflowing, since gravity can only move around gas inside a galaxy at up to a couple of hundred km/s; achieving higher speeds required energy strong enough to overcome gravity, such as supernova feedback. While it was difficult to estimate the amount of mass being carried out since these cold clumps likely comprised only a small fraction of the total outflowing material, the demonstration that galactic outflows were common at exactly the cosmic epoch when they were needed to solve the overcooling problem was a satisfying confluence of simulation predictions and observational findings.

Today, virtually every successful model of galaxy formation invokes galactic outflows driven by supernovae and young stars in order to yield a realistic-looking galaxy population. But as observations and simulations progressed, it became clear that this was only part of the story. To see why, we have to consider arguably the most important quantity in galaxy evolution: the galaxy formation efficiency.

THE (IN)EFFICIENCY OF GALAXY FORMATION

Galaxies form inside halos. Halos are straightforward to predict in a concordance cosmological model, because their dominant mass component only interacts via gravity, which we have more or less understood since the seventeenth century (with some later tweaks supplied by Einstein). But visible galaxies are another story altogether, because their evolution additionally depends on a lot of complicated gastrophysics. If we could understand how galaxies populate themselves into halos, maybe this would provide insights into the physics governing galaxy formation and evolution.

A basic way to quantify this galaxy–halo connection is via the galaxy formation efficiency. This is defined as the fraction of a halo's baryons that have formed into stars. If you remember, baryons comprise one-sixth of all cosmic mass. This means that if a halo has a given mass, the naive expectation would be that one-sixth of that mass would be in baryons. If fully one-sixth of the halo's mass were in the form of stars, the efficiency would be 100 per cent. That's quite extreme, of course; we've seen that some baryons remain in various gaseous phases, some get locked into planets or black holes, and so forth, so we expect the efficiency to be less than 100 per cent. But how much less?

This can be determined by abundance matching, which quantifies the number of stars within halos of a given mass using galaxy survey data and *N*-body simulations. One such determination by Peter Behroozi went a step further and employed data science methods to generate a self-consistent evolutionary framework using abundance matching not just at a single epoch, but across a wide range of cosmic epochs in both stellar mass and star formation rate. This state-of-the-art framework, called the Universe Machine, resulted in a galaxy formation efficiency curve shown in illustration 34.

Galaxies live along two loci: an upwards trend in efficiency at low masses and a downwards trend at high masses. The transition between the two occurs remarkably abruptly, resulting in a peak of the galaxy formation efficiency right around Milky Way-sized halos of around 10^{12} (that is, 1 trillion) solar masses. At this halo mass, the efficiency is around 20 per cent – so one-fifth of the halo's baryons have formed into stars.

The peak value of 20 per cent is already surprisingly low. If gravity and radiative cooling were only processes in forming galaxies, then a simple argument shows that the expected efficiency in this scenario should be closer to 70 per cent. Imagine some baryons falling into a halo. It turns out that it takes about 20 per cent of the age of the Universe for those baryons to reach the centre of the halo, and another 10 per cent to form into stars. So within any given halo, 30 per cent of the baryons will be 'in transit' towards forming a star – meaning the other 70 per cent should already have formed into stars. This is far higher than the observed peak of 20 per cent.

Where did the rest of the baryons go? Are they still in the halo, but somehow trapped in gaseous form? Did they fall into the galaxy but were then ejected by outflows? Did they even fall into the halo in the first place? This is such an important puzzle to solve that cosmologists have given it a name: the missing halo baryon problem.

The problem gets dramatically worse when focusing on smaller halos. For these, the galaxy formation efficiency drops

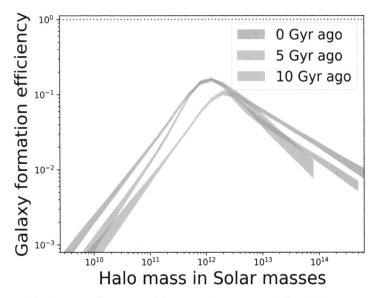

34 Galaxy formation efficiency versus halo mass, at three cosmic epochs, from the Universe Machine, shown as the shaded bands spanning the uncertainties in the measurements. The efficiency peaks at the halo mass like the Milky Way's, around 10^{12} solar masses, and drops off sharply to either side. Remarkably, this hasn't changed very much all the way back to the early Universe, 10 billion years ago, suggesting that the efficiency curve is a fundamental clue to the physics of how galaxies form and grow. Note that the scale is logarithmic, so each large marker along the x axis represents a factor of 10 in either halo mass (along the x axis) or efficiency (on the y axis).

quickly, roughly scaling with the halo mass. This means that for a dwarf galaxy living in a halo one-tenth the size of the Milky Way's – that is, 10^{11} solar masses – the efficiency is about one-tenth that of the Milky Way (just 2 per cent). Evidently, small halos are much less efficient at converting their baryons into stars. What makes galaxy formation so much less efficient in low-mass halos?

Perhaps even more surprising is the trend towards larger halo masses above the Milky Way – here, the efficiency abruptly turns around and starts getting smaller again. Why the sudden change? Is there some as yet unanticipated physics that could be kicking in to create this falling efficiency? And why does it pick out our own Milky Way as being right at the peak of galaxy formation efficiency?

The fact that the efficiency peaks exactly near our Milky Way's halo size might seem curiously anti-Copernican. However, it is anything but. Because of this peak in efficiency at this halo mass scale, it turns out that the largest fraction of stars in the Universe lives in Milky Way-sized halos. If humans were to have evolved on a random star anywhere in the entire Universe, that star is most likely to have been in a Milky Way-sized galaxy. It's like rolling a pair of dice and getting a seven – it might seem lucky, but it's the most likely outcome of any given roll. This galaxy formation efficiency curve provides an explanation, at least in a Bayesian sense, of why we live in a halo the size of the Milky Way's. Our Earth's location around an unremarkable Sun-like star in a Milky Way-sized galaxy is the very essence of Copernican.

Another remarkable fact about the galaxy formation efficiency shown in illustration 34 is that it seems to be unchanging over a large fraction of cosmic time. One could imagine that the early Universe of 10 billion years ago was a very different place to today; indeed, when we observe early galaxies, they don't look all that much like today's galaxies, being instead much more compact and forming stars much more rapidly. Yet somehow, the efficiency of converting their stars into gas isn't much different – there is still a strong peak (albeit slightly shifted upwards in halo mass) and there are still very strong drop-offs towards higher and lower halo masses at a similar rate to today. Why is the galaxy formation efficiency curve so invariant, despite a much-evolving Universe? Does this imply some fundamental regularity in the physics of forming galaxies, and if so, how does such regularity emerge?

The galaxy formation efficiency curve even relates back to the Hubble sequence: star-forming spiral disc galaxies tend to live in the rising (low-mass) part of the efficiency curve, while non-star-forming elliptical galaxies predominantly live on the falling (high-mass) part. Thus not only does the galaxy formation efficiency change at the peak, but the very nature of the galaxy population above the peak and that below the peak are markedly different. Why do the typical shapes of galaxies change across this magical peak halo mass of a trillion times the mass of the Sun?

The galaxy formation efficiency graph is one of the most influential tools for learning about why galaxies look the way they do, and by extension, why the Universe looks the way it does. Understanding why this graph has the peaky shape it does and why galaxies above and below the peak look so different goes to the very heart of the physics of galaxy formation. If a galaxy formation simulation aims to quantitatively reproduce galaxies in the real Universe, reproducing the cosmic galaxy formation efficiency graph is a good place to start.

Despite its seeming simplicity and regularity, the galaxy formation efficiency graph turned out to be a huge challenge for galaxy formation models to reproduce. It has only been within the last five to ten years that, for the first time, we have been able to simulate the formation of a realistic galaxy population with a proper efficiency curve, starting from the very early Universe. Rather remarkably, once models got that curve right, many other pieces began to fall into place. Let's see how this unfolded.

5

GALAXY FORMATION SIMULATIONS IN THE MODERN ERA

The mandate is clear: simulations of galaxy formation must include not only gravity and hydrodynamics, but galactic feedback processes that inject energy in such a way as to solve over-cooling. These feedback processes must conspire to create a galaxy formation efficiency that rises rapidly at low masses and falls rapidly at high masses, with a peak at 20 per cent around the Milky Way's halo mass. They must simultaneously produce all the attendant correlations with other Hubble sequence properties, such as morphology, star formation rate, gas content, supermassive black hole size and many more. They must do this all within a concordance cosmological framework, by growing the tiny matter fluctuations seen at the time of the CMB within representative cosmological volumes that reproduce the observed cosmic web. And they must take into account ever-improving observations enabled by ever-advancing telescopes that are continually narrowing the targets that simulations must home in on. Right, then. Piece of cake.

It is in this situation – where theory and observations (or experiments) are pushing each other forward to make new discoveries and test new ideas – that a scientific field enters a golden age. Today we are in the golden age of galaxy formation. It is akin to the golden age of particle physics in the mid-twentieth century, when ever-advancing accelerators were revealing new puzzles about the subatomic world. This led to the development of the so-called Standard Model, which today is so incredibly successful that the field of particle physics has been relegated to searching for the tiniest deviations in the hopes that they will point towards some interesting physics that is not already in the model. Likewise, cosmology similarly underwent a golden age around the turn of the millennium and is now relegated to searching for the tiniest deviations from the remarkably successful concordance cosmological model. In both areas, the theory has forged ahead of observational or experimental capabilities so far that it has become a Sisyphean challenge to find even a shred of credible evidence that subverts the dominant paradigm.

In contrast, the field of galaxy formation is in a very different place, with frequent surprises arising from rapidly advancing observational and theoretical communities. New telescopes such as the recently launched James Webb Space Telescope are giving us spectacular insights into galaxies every day. But what makes galaxy formation's golden age unique is the leading role played by numerical simulations. Particle physics and cosmology theory were driven primarily by analytic (pen-and-paper) models, but for galaxy formation we have learned that analytic theory often fails to capture the complexity of real galaxies. It is no accident that the golden age of galaxy formation coincides with the rapid rise in computing power in the Internet age.

The task of harnessing computing power to solve the problem of galaxy formation remains monumentally complicated. At the core of this mission is understanding and modelling the feedback processes that self-regulate the growth of galaxies. This is what we will cover in this penultimate chapter of the book. Welcome to the cutting edge of galaxy formation – if it seems a bit disorganized and haphazard, then this is a peek behind the curtain into how real science works before the final sanitized version gets written up in textbooks.

SIMULATING FEEDBACK: KICKS OR DUMPS?

We want to simulate the effects of stellar feedback on galaxies, but we have a fundamental problem – we don't understand how feedback works, let alone how to put it into a computer. But at the end of the day, there are two ways to prevent gas from cooling and forming into stars. The first is to prevent it from cooling and the second is to remove it from the galaxy altogether: preventive feedback and ejective feedback, respectively. In that spirit, simulators have focused on two popular subgrid models for feedback: thermal dumps and kinetic winds.

Kinetic winds take gas and simply fling it out at high speed. This removes it from the galaxy, and thus solves overcooling by lowering the amount of baryons in the galaxy. By ejecting more mass from smaller galaxies, it becomes possible to reproduce the galaxy formation efficiency curve, and thereby obtain a realistic galaxy population in our simulation. In practice, such winds are implemented via a Monte Carlo approach, which effectively involves throwing dice to determine probabilistically whether a given gas particle should be ejected. Using the Kennicutt–Schmidt relation, the code computes a probability for a gas particle to turn into a star particle. The probability of being ejected is the mass loading factor η times the probability to form into a star. The computer rolls the dice, and decides if the particle is to be formed into a star, ejected in a wind or left alone. If ejected, it is given a large velocity kick of hundreds of km/s in some random direction. To escape the dense wall of gas within the galaxy itself, hydrodynamics is shut off for a short period. When averaged over many Monte Carlo trials, the simulation properly forms some gas into stars, ejects some in an outflow and leaves the rest to be hydrodynamically evolved to the next timestep.

The second approach is thermal dumps. Here, whenever a new star particle is formed, a large amount of heat is dumped into the gas surrounding star particles to represent the thermal energy generated by supernovae and stellar winds. The heat pressurizes the surrounding gas, causing outwards expansion which leads to a galactic wind (a bit like inflating a hot-air balloon). This is much the same as the original spherical supernova feedback model envisioned by Dekel and Silk, and like that model, it works. But there is a catch – again, the dense gas within galaxies causes problems, since the high densities mean high cooling rates which quickly dissipate any heat, rendering it ineffective. In order to overcome this, simulations introduce a cooling delay, in which radiative cooling is artificially shut off for some period of time after the thermal dump, giving time for the pressure to take effect and the wind to develop.

Neither the kinetic winds nor the thermal dumps approach self-consistently generate winds from supernova energy; they are hacks introduced to produce outflows in simulations. Nonetheless, both approaches can be tuned to reproduce the galaxy formation efficiency graph, at least on the lower-mass branch. That said, most simulators are not satisfied with simply getting the right answer. We are ultimately physicists, so we would like to understand the deeper underlying reasons for how feedback works.

This is where the multi-scale approach comes into play. In recent years, zoom simulations of individual galaxies have achieved sufficiently high resolution that ad hoc tricks like shutting off cooling or hydrodynamics are not necessary. To do this properly, however, requires resolving individual stars. With the latest and greatest supercomputers, this is now beginning to be possible within simulations of single small dwarf galaxies, such as in the GRIFFIN Project led by Thorsten Naab and collaborators.

But zooms of larger galaxies like the Milky Way remain well out of reach of modern computing power, let alone full cosmological volumes.

Zoom simulations have nonetheless enjoyed some encouraging successes at generating galactic winds more consistently from the stars formed within the simulation, even if not quite resolving individual stars. Current examples include the FIRE and NIHAO zoom suites, among others. These use state-of-the-art hydrodynamics techniques, and include a host of detailed physical processes for stellar evolution and supernovae. All have been successful at producing individual galaxies with the right galaxy formation efficiency. These zooms also make direct predictions for the mass loading factor, which can be implemented directly into cosmological models in a truly multi-scale approach to modelling feedback.

By combining zooms and cosmological simulations, it's possible that we are converging on a more holistic understanding of how stellar feedback works. We're not there yet, because in a sense we now have too many successful models, which all implement feedback differently – and surely they can't all be right. But simulations are now getting realistic enough that perhaps they can give insights into what such feedback implies for how galaxies form and grow. In the last decade, this has indeed happened. The result was that yet again galaxy formation simulations have driven a paradigm shift in our conception of how galaxies form and evolve.

THE BARYON CYCLE

Cosmological galaxy formation simulations can now produce realistic galaxy populations. That is an accomplishment, but a cynic might say (not entirely unjustifiably) that we have simply tuned a bunch of parameters to get the answer we want, which, of course, isn't science. What have we learned from this?

The answer, in a broader sense, is emergent phenomena. Emergent phenomena are behaviours arising from the interplay of numerous complex processes that are not anticipatable by considering each process individually. Cosmological simulations connect a huge range of physical phenomena, from the growth of the cosmic web on large scales to the formation of stars and black holes on small scales. Perhaps the most remarkable result

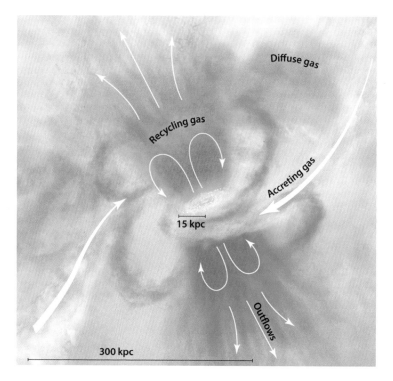

35 Schematic drawing of the baryon cycle operating around a disc galaxy, showing the ways that a star-forming galaxy interacts with its environment. Inflows provide fuel to grow galaxies, outflows remove fuel to retard growth, and recycling provides an additional source of fuel. The amount and interplay of these baryon cycling processes determine what a galaxy looks like and how it evolves over time.

from these simulations is how closely these processes, which span enormous scales, are interdependent, such as the growth of stars deep within galaxies relying on the gaseous fuel supplied by the cosmic web. This is why simulations are truly greater than the sum of their parts.

From galaxy formation simulations, arguably the most interesting and far-reaching emergent phenomenon is the baryon cycle (illus. 35). The prevalence of galactic winds makes it evident that the journey of gas from deep intergalactic space into galaxies is not a one-way trip; gas flows not only into galaxies, but out of them. In current simulations the cosmic-averaged mass loading factor is well above one, which means far more gas is being ejected from galaxies than is forming into stars. Epistemologically, this is a game-changer. If simulations are correct, visible stars are nothing but the frosting on a massive circulation of matter into and out of galaxies; it just so happens that it's the frosting that shines most brightly.

The great thing about simulations is that they don't just provide snapshots, they provide the whole film. One way to view this film is to track individual particles as they move around. When doing this in hydrodynamic simulations, it was found that gas cycled many times through a galaxy over its lifetime before forming into a star. Such wind recycling is far from a negligible curiosity; for our own Milky Way, simulations estimated that the typical star today was formed from material that had been in a galaxy long ago but had been expelled at least once into intergalactic space before being pulled back in. Matter can even be expelled from one galaxy and fall into another, in a process called intergalactic transfer; FIRE zoom simulations suggest that up to half the baryons within our Milky Way today may have originated from another galaxy.

It seems that galaxies, far from being the isolated 'island universes' envisioned by Kant, are part of a dynamic cosmic ecosystem within which they grow and evolve over time. The particular evolution of an individual galaxy like our Milky Way is some combination of 'nature', namely the strength of the initial density perturbation from which it arose, and 'nurture', its interactions with its cosmic ecosystem via the baryon cycle. This is a very different view to the canonical idea of isolated halos forming their own galaxies and only occasionally interacting with other halos via mergers. Instead, modern galaxy formation simulations suggest that what we view as an isolated galaxy floating in an ocean of emptiness is akin to a city skyline at night, a veneer of majestic serenity atop a chaotic bustle of energy and motion.

I SAW A LINE THAT WASN'T THERE

The baryon cycle view of galaxy evolution is a progressive twenty-first-century story, complete with recycling and ecosystems. But does it have anything to do with reality? As the saying goes, trust but verify. There would be no better verification than to directly observe the baryon cycle in action in the real Universe.

This is immensely difficult. The inflows and outflows characterizing the baryon cycle are happening in the tenuous surrounding gas, which has come to be known as the circum-galactic medium (CGM). The tenuous gas moving in and out of galaxies doesn't shine very brightly, and even if it did, how does one tell what gas is inflowing, what is outflowing and what is ambient, and how does one measure the total amount of mass and energy in each component? Quantifying the baryon cycle strains the limits of current telescope technology.

As it happens, instead of looking for what's there, there is more information to be gained by looking for what's *not* there. Suppose, for argument's sake, that someone placed a bright torch behind the CGM of a galaxy. And suppose you knew the spectrum of the torch, where a spectrum is the intensity of light coming out at each wavelength (or colour). As that light passes through the CGM of the foreground galaxy, it will encounter atoms of gas – hydrogen and helium, but also heavier elements such as carbon, oxygen and silicon – which were produced inside stars and carried into the CGM via galactic winds.

Elements leave fingerprints in the torch's spectrum. We discussed this earlier when we talked about radiative cooling via electrons moving between energy levels within a hydrogen atom. Like hydrogen, each element in the periodic table has its own pattern of energy levels. If an electron is hit by a photon with precisely the right energy, it can swallow that photon's energy and jump to a higher energy level. As it does so, it effectively takes a bite out of the light at a specific energy, which itself corresponds to a specific wavelength. This gives rise to an absorption line.

Like fingerprints, the unique pattern of each element's absorption lines allows astronomers to tell which elements are present in the CGM, and the size of the bites indicates how much of each element is there. This is how, by seeing what is missing, we can tell what is there (illus. 36), even if what is there is not shining.

The main requirement for this is that there must be a bright torch behind the CGM of a galaxy. Conveniently, nature has placed many bright torches all over the sky. We call these quasars. Quasars are actually supermassive black holes that are currently undergoing a feeding frenzy, causing a huge amount of high-energy light to emerge from the black hole's accretion disc. They are amazing objects in their own right, as we will discuss in a bit, but for the purposes of absorption line spectra, all we care about now is that they provide a bright background light source that pierces a galaxy's CGM.

Nature doesn't make it too easy for us, though. For one thing, quasars aren't very common, so it is rare to find one aligned directly behind a galaxy's CGM. Moreover, many of the strongest finger-print lines for common elements such as hydrogen, silicon and carbon lie in the ultraviolet, which requires sending telescopes into space in order to get above the Earth's UV-blocking atmos-phere. Even if the patterns can be decoded, individual absorbers only trace a single phase of that element, so obtaining a full census requires uncertain extrapolations to account for all the other gas phases. Furthermore, there can be many absorbers within a single CGM, some of which are inflowing, some outflowing and some ambient; it is not usually obvious which is which. And even if all of these can be sorted, absorption lines only provide one-dimensional skewers through the CGM, and thus the skewers contain much less information than two-dimensional images or three-dimensional simulations. While having only some informa-tion is better than having none at all, CGM absorption lines by themselves are notoriously difficult to connect to baryon cycling.

This is why simulations have emerged as an invaluable com-panion to CGM absorption line data. In a simulation, all the infor-mation is there – the amount of hydrogen and heavier elements, what phase they are in and how they are moving around. It is straightforward to place a pretend torch behind the simulation volume and generate a mock-quasar spectrum as if the simulation were a patch of the real Universe. The resulting spectrum con-tains all the absorbers in various elements that would show up if this were a real galaxy's CGM with an actual quasar behind it.

Comparing the statistics of such mock spectra from simulations versus Hubble's real quasar spectra gives us some idea whether the simulation resembles reality. If so, this allows us to learn about the baryon cycle by proxy, taking advantage of a simulation's full three-dimensional information.

Recent galaxy formation simulations do a creditable job of reproducing CGM absorption line data – but only when they include strong galactic winds. This is because a lot of heavy elements are seen in the CGM, and the only way these elements can be transported far away from the stars where they are created is via strong galactic winds. Subgrid feedback models are usually tuned to match galaxy properties (predominantly to solve over-cooling), so it is a non-trivial and encouraging success that these same simulations are realistic in a very different regime of the Universe, namely the CGM. While this doesn't constitute a direct detection of baryon cycling, it nonetheless gives us some confidence that the baryon cycle is not just a figment of a computer's imagination.

THE CASE OF THE RED AND DEAD GALAXY

Things are looking up in the galaxy formation game. We've managed to solve overcooling, incorporate stellar feedback processes, understand CGM absorption lines and come up with a holistic baryon cycling view of galaxy evolution within a cosmic eco-system. Yet through all of this, we've managed to skirt around a rather large elephant in the room: what about the other side of the galaxy formation efficiency curve?

Galactic outflows and the baryon cycle yield the rise in efficiency from the lowest masses up to a peak at the magic halo mass of a trillion solar masses. But going to even higher halo masses, the trend of increasing efficiency abruptly reverses – galaxies inside more massive halos turn out to be less efficient at forming stars. What is happening at this magic halo mass of about a trillion (10^{12}) solar masses? Why the abrupt change? Moreover, why do galaxies of the dominant Hubble type change from spiral discs to ellipticals (illus. 37)? Surely this change in galaxy type must also be an important clue.

Let's don our Sherlock Holmes caps again and follow the energy. If galaxy formation is less efficient, there must be some additional source of energy at higher halo masses. But here is where it gets confusing: these massive galaxies tend to be quenched – that is, they haven't formed any new stars for a very long time. Up until now, we've invoked the energy from young massive stars,

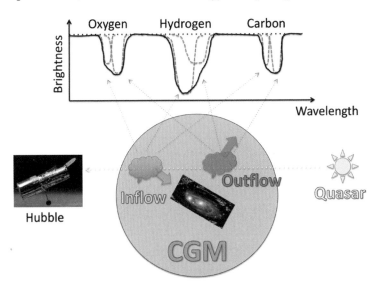

36 CGM absorption lines. A background quasar shines through a galaxy's CGM with inflowing and outflowing gas. The different elements, such as hydrogen, carbon and oxygen, produce absorption lines at specific wavelengths, resulting in a final spectrum (shown in black) seen by using, for example, the Hubble Space Telescope. The pattern of absorptions allows us to decipher which elements are present and the depth tells us how much of each element is there.

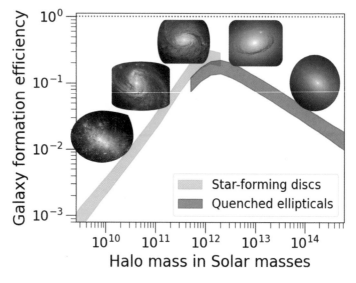

37   Galaxy formation efficiency, split into star-forming discs (cyan) and quenched ellipticals (magenta). The abrupt transition in efficiency is accompanied by a similarly abrupt transition from primarily star-forming discs to predominantly quenched ellipticals. The Milky Way's halo lies at the peak of this curve, but our Milky Way galaxy is unusual because only about one-third of the halos at this transition mass contain star-forming discs; the majority contain ellipticals.

but in elliptical galaxies there are no young stars. To make matters worse, the energy input has to be truly enormous in order not only to regulate star formation, but to cease it altogether! What source could possibly provide such a stupendous amount of energy?

   Let's look for clues in the differences between star-forming and quenched galaxies. Here are some key points of difference: quenched galaxies (i) are usually elliptical in shape; (ii) tend to live in denser nodes of the cosmic web; (iii) are often surrounded by lots of hot gas, visible in the X-rays; and (iv) tend to have huge central supermassive black holes, billions of times the mass of the Sun. Can any of these traits yield a massive amount of energy that is powerful enough to quench star formation altogether?

The elliptical shape might be a hint of a galaxy merger. A merger is a very energetic event, and we often see starbursts and strong winds associated with merging galaxies. Mergers turn spirals into ellipticals (illus. 38), so this hangs together nicely with the idea that massive galaxies seem to be elliptical in shape. If the merger could concurrently provide enough energy to blow out all the galaxy's gas, then star formation could cease. For a while, this was the leading idea for how galaxies become quenched while also having lower efficiency.

While the idea looked promising on paper, galaxy formation simulations showed that it doesn't really work in a cosmological setting. Mergers can indeed evacuate gas, but only temporarily;

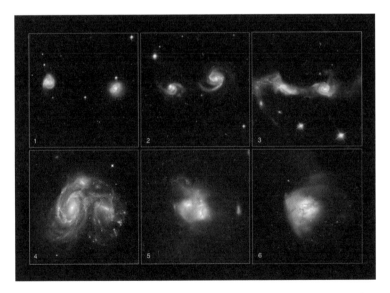

38 This montage of Hubble Space Telescope images shows galaxies seen at various stages of merging. Two spiral galaxies approach each other, and the strong tidal forces disrupt the spiral structure, resulting in a train wreck often accompanied by a burst of star formation. After the starburst fades, the remnant galaxy is a featureless elliptical that has no spiral structure. This spheroid can later regrow a disc around it by accreting more gas, but it will retain a large central bulge of older stars.

after the big brouhaha, eventually the galaxy settles back down, gas begins to be pulled back in towards the galaxy and the object starts forming stars again after a billion years or so. This doesn't fit in with observed quenched galaxies that have not been forming stars for the last 8–10 billion years. It appears that mergers alone are not energetic enough to shut off star formation permanently.

What about the fact that quenched massive ellipticals tend to live at the densest nodes of the cosmic web surrounded by hot gas? Promisingly, these regions have so much gravity that they draw in matter supersonically, causing the gas to undergo violent shock heating. Unfortunately, such shock heating occurs in the outskirts of the halo, far from the galaxy. Getting that heat to the halo's centre in order to stop star formation requires a form of energy transport, such as conduction or convection. But magnetohydrodynamic simulations suggest that such processes are far too slow and inefficient, even under optimistic assumptions. While location in the cosmic web might help quench galaxies a bit, it cannot by itself quench them altogether.

At this point, the well-known Arthur Conan Doyle saying comes to mind: once you eliminate the impossible, whatever remains, however improbable, must be the truth. And so our gaze falls upon that devourer of stars, that behemoth of doom, that place where God divided by zero: the central supermassive black hole.

WHEN YOU'RE A JET, YOU'RE A JET

The good news is that black holes are capable of accelerating material close to the speed of light, and there are lots of energetic processes that can happen when you accelerate a plasma up to ludicrous speed. The bad news is, black holes are famous for not

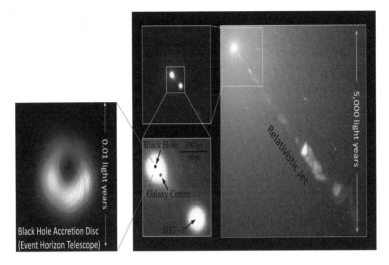

39 The black hole in M87 and its jet. The image on the left shows the black hole accretion disc imaged by the Event Horizon Telescope. This lies near the centre of M87, and the lower left panel of the right image also shows a bright knot (HST-1) moving outwards in the jet. The large-scale image shows the relativistic jet emerging from the core of the galaxy, seen in the optical with HST. The jet can be traced significantly further out into intergalactic space at radio wavelengths (not shown).

letting anything escape, even light. So how are we supposed to extract any energy from a black hole?

Sometimes, it pays to look at an actual galaxy to see what might be going on. Let's look at a galaxy called Messier 87 (M87), which is a large elliptical galaxy that lives at the centre of the nearby Virgo Cluster. In one of the more stunning recent results in astronomy, M87's billion-solar-mass central black hole was directly imaged by the Event Horizon Telescope (EHT), a network of telescopes spread across the globe.

M87 has another remarkable feature: a jet. A jet is a narrow stream of hot ionizing gas that shoots out at speeds approaching the speed of light from deep within the galaxy into intergalactic space. Jets have long been suspected to arise from central black

40 Hercules A is an elliptical galaxy with spectacular bipolar jets driving superheated ionized gas at relativistic speeds into its surrounding circum-galactic medium. The optical image showing the starlight from galaxies is overlaid by a radio image (in magenta) showing synchrotron emission in lobes where the jets interact with circum-galactic gas (which is not visible in emission in these bands). These interactions heat the CGM, preventing it from cooling onto the galaxy and keeping the galaxy quenched.

holes, and thanks to the EHT image, it was conclusively confirmed that M87's jet indeed originates right at its black hole (illus. 39).

Jets carry a lot of energy out into intergalactic space. This can be seen as radio synchrotron emission released when it runs into the surrounding circum-galactic gas; a spectacular example of this is Hercules A (illus. 40). The radio lobes are filled with extremely hot gas. Over time, this heat will dissipate into the surrounding CGM, countering radiative cooling. The hot CGM thus quenches the galaxy by slowly starving it of fuel for forming new stars. But the supermassive black hole is necessary for providing the energy to do so.

The hot CGM can be observed in X-ray emission, particularly in enormous galaxy clusters that contain thousands of galaxies trapped within a huge dark matter halo. An example is MS0735, where illustration 41 shows a composite among three images taken with different telescopes: the X-ray emission, shown in blue and traced by NASA's Chandra X-ray Observatory, comes from hot gas; the yellow shows galaxies imaged by the Hubble Space Telescope; the red shows a Very Large Array radio telescope image of emission from jets emanating from the central galaxy's supermassive black hole. These jets are how the massive central cluster galaxy becomes red and dead, by adding heat and preventing the surrounding hot gas from cooling and providing fuel for star formation. MS0735 is a smoking gun example of how this cosmic galactic suicide plays out.

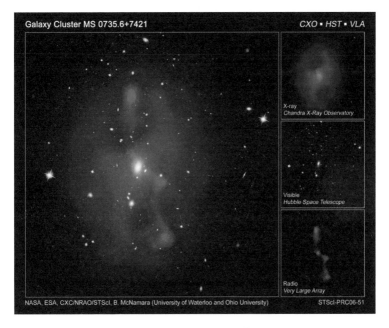

41   Hot gas seen in X-ray emission being heated by radio jets from a supermassive black hole in a galaxy cluster. This composite image shows the stars in yellow. From deep within the central massive galaxy, a radio jet is launched from a supermassive black hole. It keeps the X-ray gas (shown in blue) hot, preventing star formation and ultimately making the galaxy 'red and dead'.

Humans haven't yet come close to solving the mystery of how nature quenches galaxies. Stitching together the entire process of generating a jet from the accretion disc of the black hole out to millions of light years into diffuse intergalactic space is an enormous computational challenge that isn't close to being solved. But with growing evidence that jets from black holes must have something to do with killing galaxies and reducing their efficiency, we're going to have to account for feedback from black holes in our simulation, whether or not we have any idea of how it actually works.

BLACK HOLES ARE MESSY EATERS

Of all the gas and stars that make it down into the accretion disc surrounding the black hole, only about 5–10 per cent ends up going into the black hole. The rest is spat out, in a variety of highly energetic ways, including in the form of high-energy X-ray photons, radiatively driven winds and the aforementioned jets. All of these processes are occurring on scales of the black hole's accretion disc, which is smaller than a light year across. Cosmological-scale simulations, with spatial resolutions of hundreds or thousands of light years, have no hope of directly modelling these processes. So, once again, we must resort to subgrid models, both to grow the black hole and to account for its feedback energy.

This is the bleeding edge of modern galaxy formation simulations. The approaches used are similar to the ones used for star formation feedback, that is, thermal and kinetic feedback – but on steroids. In the thermal feedback approach, gas around the black hole is superheated to insane temperatures, up to a billion Kelvin, representing the massive energy deposited into the surrounding galaxies. Other simulators prefer the kinetic approach, since it more closely resembles real jets, by ejecting material at many thousands of km/s (as opposed to the hundreds of km/s speeds of star-formation-driven winds). These sorts of extreme feedback implementations stress the limits of hydrodynamics methodologies and computational capabilities. They can cause all kinds of numerical instabilities. But it is not by choice that black hole feedback is included in all currently successful galaxy formation models; it's by necessity.

The tremendous energy from black hole feedback can have dramatic effects on galaxies and their circum-galactic medium. In some modern simulations, the CGM just above the magic turn-over mass is almost entirely evacuated of gas. Illustration 42 shows the fraction of baryons in various phases within halos, as a function of halo mass in my group's SIMBA cosmological hydro-dynamic simulations. The difference between the left and right panels is that the left includes black hole jet feedback, while the right doesn't. You can see that this makes a huge difference, particularly above 10^{12} solar masses. The dramatic lowering of CGM gas, particularly cold gas, kills the new star formation in the galaxy and causes the stellar fractions to remain low even up to the largest halo masses (in contrast to the no-jet case). This illustrates how using simulations as numerical experiments can give insights into the important physical processes governing the formation and evolution of galaxies.

Black hole feedback models in simulations are in their infancy. The parameters are crude and are constrained mostly by the 'just tune, baby' approach, adjusted to match observed relations such as the correlation between black hole mass and galaxy mass, and the turnover in the galaxy formation efficiency. They work, but it is not entirely clear how much physical insight is contained in those models, especially since different simulations currently use

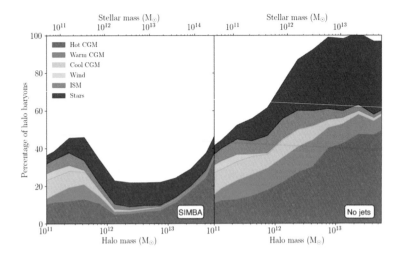

42   Baryon fraction in halos in the SIMBA simulation (left panel). This shows the mass of baryons in various CGM phases in each coloured band, divided by the expected baryon content of the halo (namely, one-sixth of the halo mass), as a function of halo mass (with approximate corresponding stellar mass along the top). The overall baryon fractions are always under 50 per cent, showing that more than half the expected halo baryons have been evacuated by feedback. The stars are the dark blue band, which is fattest in the middle, indicating the peak efficiency there. The hot gas increases in large halos, while cooler gas phases dominate the CGM of low-mass halos. The right panel shows the exact same simulation with the black hole jet feedback turned off (but stellar feedback still on). In more massive halos, the black hole jets are largely responsible for evacuating the halo's baryons; stellar feedback alone still manages to evacuate the halo somewhat at low masses.

very different subgrid models for black hole feedback yet all can match galaxy observations similarly well.

Today, all successful galaxy formation models invoke star formation feedback below the turnover mass, combined with black hole feedback above the turnover mass, in order to get the galaxy formation efficiency curve correct. While there are huge uncertainties in the underlying physics behind each, there is growing hope that insights from detailed zoom simulations on ever more powerful computers will eventually bridge the dynamic range gap with cosmological simulations, and complete the story of how galaxies and their cosmic ecosystems originated from the Big Bang.

ARE WE THERE YET?

With subgrid models for star formation, galactic outflows, black hole growth and black hole feedback in place, current cosmological simulations can at long last reproduce a fully realistic-looking Hubble sequence of galaxies, with all their attendant properties, starting from primordial matter fluctuations arising out of the inflationary epoch. This has only happened in the last few years, and represents a landmark achievement in galaxy formation simulations.

Does this mean we are done? Far from it. In many ways, the process is just beginning. The main problem is that all these subgrid models are, for the most part, crude and unsatisfactory. They are not genuine models of physics, but rather cartoon parameterizations thereof. Hence the main current direction in galaxy formation theory is to understand the subgrid physics that seems to be so critical for setting the properties of galaxies. The hope is that a better understanding of how star formation happens, how it drives outflows, and how black holes grow and release energy will provide insights into ways to increase the realism of cosmological simulations.

Meanwhile, the cosmologists are pushing the galaxy formation community in the opposite direction, towards larger scales. The impact of these feedback processes on cosmic ecosystems appears to be surprisingly widespread – for instance, if it's true that the majority of baryons in massive halos have been ejected out of galaxies, as some simulations suggest, then this strongly impacts cosmological measurements of dark matter and dark

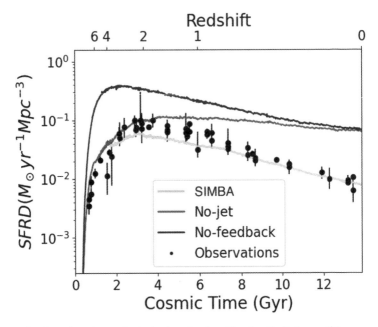

43 Cosmic star formation rate density (SFRD) as a function of time from the Big Bang until the present. Circa 2000, simulations did not include feedback, like the top line shown, and formed far too many stars compared to observations (black points); this is the overcooling problem. Adding stellar feedback but as yet no black hole feedback (no-jet line), such as simulations of circa 2010, suppressed early star formation and solved the too-bulgy galaxy problem, but still failed in the latter half of cosmic time because it did not lower the efficiency in massive galaxies. Finally, we reach 2020: the lightest line shows our SIMBA simulation, which includes both stellar and black hole feedback. This matches SFRD observations well, and produces a galaxy population in reasonable accord with the real Universe. This graph illustrates the remarkable progress that galaxy formation simulations have made over the last twenty years towards producing a realistic universe of galaxies.

energy, since there is more baryonic mass between galaxies than previously thought. It would be disastrous if new facilities such as the Large Scale Synoptic Telescope, ESA's Euclid mission and NASA's Nancy Roman Space Telescope, constructed to the tune of billions of U.S. dollars, successfully mapped the cosmic web in unprecedented detail, but their measurements of dark matter and dark energy were flawed because they fail to account for how feedback processes have moved mass around the cosmos. Hence cosmologists want galaxy formation simulations to run ever-larger volumes, in order to probe the scale of the cosmic web relevant for their experiments.

With galaxy formation simulations now producing realistic galaxies, this sets us on course for using these simulations as 'numerical experiments', with which we can study the impact of specific physical processes on galaxies. This is a huge help, because we can't conduct experiments on the real Universe – unlike chemists in a laboratory – and so we can't very well tell a galaxy to turn off its black hole or stop having supernovae. But we can do that in a simulation.

An example of a numerical experiment is shown in illustration 43. In my group's SIMBA simulation (indicated by the faint grey line), which includes star formation and black hole feedback, we can nicely reproduce the observed rate of stars forming in the cosmos over all of cosmic time (the black points). We then ran a model where we turned off the black hole jet feedback just to see how different the universe would look. This model looks fine for the first couple of billion years, but then it goes off the rails, because massive galaxies are not quenched, meaning their efficiency is too high. Finally, we turn off all feedback (black hole and stellar), which is the darkest line. Here, way too many stars are formed at all cosmic epochs, showing the classic overcooling problem.

Illustration 43 encapsulates the dramatic progress in cosmological galaxy formation simulations over the last twenty years. The top line, with no feedback, was what simulations were like circa 2000. The inclusion of supernova feedback, appropriately tuned, was able to solve early overcooling, but could not produce quenched elliptical galaxies; this was state of the art circa 2010.

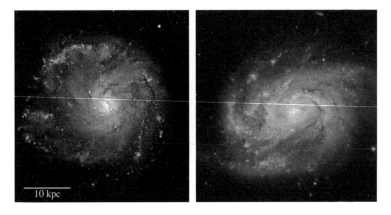

44 Spot the fake! One of these is a real galaxy and the other a simulated galaxy from a zoom simulation by the FIRE project: which is which? The one on the left is a Milky Way-sized galaxy from a FIRE zoom simulation called Thelma, while the one on the right is an actual galaxy, NGC 1803. These are not intended to be the same galaxy; they illustrate how realistic today's zoom simulations can look.

Finally, the lightest line represents a modern cosmological simulation, SIMBA, that both solves overcooling and produces a turnover in the galaxy formation efficiency curve. This graph may look like just a small shift of some lines, but achieving this required enormous amounts of new input physics, improved computational algorithms, tedious trial-and-error, painful dead ends and consumed caffeinated beverages.

Perhaps the most lasting lesson from this whole enterprise is how interconnected the Universe is. Processes occurring on scales well below a light year, such as star and black hole formation, can have a major impact on matter distributed on scales of millions of light years throughout the cosmic web. Such an intimate connection, spanning such a large range of scales, is virtually unprecedented in science, and galaxy formation sits right smack in the middle of this. This, more than anything, illustrates why solving galaxy formation is truly the lynchpin for understanding why the Universe looks the way it does.

6

THE FUTURE IS NOW

A rmed with the concordance cosmology and ever-increasing computing power, the pace of progress in cosmological simulations has been astounding. One could rightly ask, when will it be finished? When can we say that we've answered the question of why the Universe looks the way it does?

In a pedantic sense, the answer is never. There will always be more to discover about the Universe, about the Milky Way, about the formation of stars like the Sun and planets like Earth and all the other mysterious objects that abound in the cosmos. One of the many humbling aspects of studying astrophysics is the stark realization of how little humans know about the Universe, and that the unknown unknowns almost certainly dwarf the known unknowns. But this is the nature of science – it is a quest that is never complete; the enjoyment lies in making the journey rather than in arriving at the destination.

PRECISION GALAXY FORMATION

More practically, we could potentially declare the area of galaxy formation as solved if we achieve a state of the field such as that currently seen in particle physics or cosmology. Cosmology's transformation over the last several decades provides a template of sorts. Cosmology went from being an intellectual free-for-all to a precision science, with a well-established concordance model within which the basic parameters of the cosmos have been identified and measured to a few per cent accuracy or better. Today, it is said that we are in the era of precision cosmology.

Analogously, solving galaxy formation is tantamount to making it into a precision science, where all the main physical ingredients have been identified and the governing parameters have been measured to a a precision level similar to that of cosmological parameters today. We want to develop a 'standard model' for how galaxies form and evolve in our Universe, backed by a deep understanding derived from first principles of physics. We want that standard model to be able to withstand any new set of observations from the latest facilities in the way that the concordance cosmological model or the Standard Model of particle physics do.

As of now, galaxy formation remains far from this. It is not even entirely clear what constitutes a set of governing parameters in galaxy formation theory, let alone being able to constrain their values to per cent level. Frustratingly, the complexity only seems to increase when we look towards smaller and smaller scales. How can we possibly understand how galaxies form and evolve if we don't understand how their stars, black holes, dust and all their other constituents form and evolve?

If I had to place my bets on one approach that could yield a precision model of galaxy formation, it would be on the baryon cycling paradigm. In baryon cycling, galaxy growth is primarily governed by gas flowing in, some of which forms into stars and some of which is ejected in outflows, with some of those outflows raining back on galaxies via wind recycling. This constitutes a well-defined set of physical processes that can be parameterized in some way. Even if we don't understand the physical origin of star formation and outflow driving, that doesn't invalidate the macroscopic approach. The situation is analogous in cosmology

– we don't understand the nature of dark matter and dark energy, yet we can still construct a model that describes our Universe and its evolution exceptionally well on very large scales. Is it likewise possible that we could measure baryon cycling to sufficient precision, either directly or indirectly, to construct a model that yields the galaxy population as we see it? This might not be as hopeless as it seems.

AN INTUITIVE MODEL FOR BARYON CYCLING

How would we go about constructing a simple, intuitive model for baryon cycling? A good place to start, rather like the follow-the-energy mantra we had earlier, is to follow the mass. Mass flows into galaxies, some fraction goes into stars and some into winds, and some portion of outflows returns as inflow. We merely have to parameterize how much mass is participating in each stage of this baryon cycle, and then constrain those parameters. Let's walk through how this might work.

The starting point is, as usual, a dark matter halo. Given our Concordance ΛCDM cosmology, N-body simulations can predict the growth history of individual dark matter halos very accurately. Let's make our first ansatz, which seems fairly reasonable: the inflow of baryons into a halo is one-sixth that of the total matter, following the cosmic baryon-to-total-mass ratio. With that, we can now get a baryonic inflow rate into a halo.

From there, the baryons can do one of two things: fall down into the centre of the galaxy via radiative cooling or remain hot and held up in the halo. So we come to our first parameter, quantifying preventive feedback: the fraction of baryons that is able to reach the galaxy. Let's denote this parameter by the Greek letter zeta (ζ). This is our first baryon cycling parameter.

For each baryon that makes it into the galaxy, again one of two things can happen: either it forms into a star or is ejected in an outflow. We have already encountered a quantity that is the ratio between these two quantities: the mass loading factor, η. That's our second baryon cycling parameter.

Finally, that ejected matter can fall back into a galaxy, adding to the inflow associated with dark matter. To parameterize this we can define a recycling time, t_{recyc}, which governs how quickly ejected material returns to the galaxy.

None of these baryon cycling parameters {ζ, η, t_{recyc}} needs to be a single number for all galaxies at all times. They could depend on, say, the mass of the galaxy or halo, or the cosmic epoch. These dependencies are not known a priori, but modern galaxy formation simulations give us some sense of how they must work. For instance, simulations suggest that ejective feedback (η) must be stronger in lower-mass galaxies in order to get lower galaxy efficiencies in smaller halos. Meanwhile, preventive feedback (ζ) must quickly become stronger at higher masses in order to quench galaxies and get the abrupt turnover in the efficiency graph. Simulations also make predictions for the wind recycling time t_{recyc}, but such predictions vary greatly between models. So we at least have some guidance as to how to choose these parameters, even if the exact values are not known.

The baryon cycling equation is then a simple one tracking the mass among these various processes:

Inflow + recycling = star formation + outflow

This represents the conservation of mass; the amount of mass falling in equals the amount of mass ejected out plus the stuff formed into stars. This simple equation relates all the baryon cycling parameters.

It connects to cosmology because the inflow rate for any halo depends on the cosmological parameters that set the growth rate of halos in the cosmic web. The baryon cycling equation gives a quantitative model for the amount of star formation (and hence stars) in a galaxy over time.

Various galaxy formation theorists including myself have developed such baryon cycling models, and they are surprisingly successful at describing the growth in stars of the overall galaxy population. With just a handful of free parameters constrained to match galaxy survey data, such models can reproduce basic galaxy demographics such as the galaxy formation efficiency, and even the galaxy's gas and heavy element content. Although this might seem like a trivial result (with enough parameters, any model can fit any data), it is remarkable that such a simple framework is so effective at generating a sensible population of galaxies; one can think of a great many complexities associated with galaxies that are not accounted for in this simple baryon cycle scenario (such as, for instance, galaxy merging), yet none of them seem to fatally break the model.

While encouraging, these models are extremely simplistic. They don't, for instance, predict galaxy morphologies or their black holes. They don't come anywhere close to reproducing the full range of observed properties of galaxies that we can access with our latest telescopes spanning the electromagnetic spectrum. So they are not a replacement for full hydrodynamic simulations by any means. Their main advantage is that they are much more intuitive than a bunch of particles in a virtual box. With such baryon cycling models, it is much easier to understand which physical processes govern which aspects of galaxy growth.

A truly successful physical theory should be expressible intuitively, even if that intuition doesn't explain every last detail. Baryon cycling is currently the best angle we have towards an intuitive model for how galaxies form and grow over time. Undoubtedly it isn't the whole truth. Science is not so much about finding the truth, though, as about finding the best possible explanation for the world around us. Baryon cycling models may be the first step towards building a simple and intuitive model of precision galaxy formation.

GALAXY FORMATION SIMULATIONS, VISION: 2100

Perhaps as much as any field in science, astronomy is at the mercy of technological development in order to make progress. Indeed, this is a significant reason why governments decide to fund astronomy. Astronomy not only captivates the public's imagination, but investments in the field push forward the limits of technology. Perhaps the best example of this is the charge coupled device (CCD), an integral part of all modern mobile phones and digital cameras, and a technology developed by astronomers many decades ago in order to take more accurate pictures of the night sky. The benefit to the general public associated with developments in astronomy's technology may not be immediately obvious, but the long-term payoff is undeniable.

Even within astronomy, numerical simulations are particularly reliant on technological advancement. Moore's Law has been a driving force behind the scenes in the progress of cosmological simulations over the last few decades, but computer chips can only be made so small and processor speeds so fast before running up against the limits of physics. Lately, progress in computing speed has been driven less by increases in individual chip speeds and more by specialized technologies such as graphics processor units (GPU) that use existing chips in clever ways. Unfortunately, such specialized technologies are not as flexible to use, and it is a continuing challenge in astronomy to adapt the codes we use to model the Universe onto the latest technologies that the computing industry wants to drive forward. Such bleeding-edge technologies are often optimized for applications such as database management or gaming – that is, uses that are a far cry from astrophysics. The result is that we numerical astronomers are constantly working to trim our sails to the tech industry's winds.

Massive parallelization has been the driver of the Internet era. The idea is that rather than using insanely expensive chips with cutting-edge speed, we should use cheaper hardware and gain power in numbers. The first computing cluster was a machine called Beowulf, built in 1994 at NASA by stringing together 52 ordinary servers. For a time such systems were referred to as Beowulf clusters. By using commodity hardware and free open-source software such as Linux, the cost per computing cycle is kept low. Today, such computing clusters are everywhere, and are the engine driving the Internet.

Getting astrophysical simulation codes to operate in a cluster environment is no easy feat. If one has a million completely independent tasks, like Google database searches, it is easy to put one on each server. But in the case of a cosmological simulation, a particle in one corner of the simulation interacts via gravity with a particle in the other corner, even if the interaction is weak; nothing is isolated or independent. When computing gravity, therefore, the first node has to send a request to the second node over some Ethernet-like connection, and the second node then has to acknowledge that request, look up the desired information and send it back to the first node. An entirely new computer language, message passing interface (MPI), was invented in order for computers to process such requests. Simulation codes that run on a computer cluster are thus programmed not only in low-level languages such as C or C++ or FORTRAN, but in MPI, with specific message passing requests arranged to ensure all the forces are computed properly.

As you might imagine, because of the cost in time of sending information via the Ethernet, parallel computing is significantly slower than doing a computation on a single server. But it is the only way to run large state-of-the-art simulations. As a naive PhD student without any formal computer science training, I developed the world's first MPI-parallel SPH-based galaxy formation code. However, it was quickly superseded by Springel's Gadget code, which solved the same equations but used computer science techniques to reorganize the calculations in an efficient manner for parallel environments, and thus was much faster. This was an early example of how ideas from computer science were instrumental in fashioning leading-edge simulations.

Twenty years later, Gadget is now becoming somewhat archaic, as many improvements have occurred in computing technologies in the intervening years. While Gadget continues to be updated, other groups have taken a fresh approach. For instance, Durham University's Institute for Computational Cosmology has partnered with computer scientists to completely redesign the structure of a cosmological simulation code to work more efficiently on modern architectures via task-based parallelism. Their SWIFT code can reportedly do the same simulation on the same computer as Gadget in just a fraction of the time. This shows that there is still much room for growth in galaxy formation simulations simply by using existing technologies in smarter ways. Yet even SWIFT does not take advantage of the massive computing power in GPUs, the special-purpose cards that accelerate gaming to drive their stunningly lifelike graphics. Perhaps the next great cosmological code advance will find a way to take advantage of GPUs, which so far in astrophysics have only been employed in specialized circumstances that are amenable to its restrictive architecture.

In the computing industry, the great hope for the future is quantum computing. The basic idea is that a collection of quantum states has, under certain conditions, properties that allow it to reconfigure itself via the laws of quantum mechanics. These reconfigurations can then be envisioned like a normal transistor switching between zero and one, except instantaneously and on a massive scale. A small class of computational problems lend themselves well to this set-up, and allow for those algorithms to be solved, in a sense, at the speed of nature. But quantum computers remain very far from being general-purpose numerical solvers that can run ordinary apps like a web browser. Nonetheless, if quantum computing takes off, it is certain that numerical cosmologists will get to work on adapting this technology for galaxy formation science. The job of simulators has increasingly shifted towards not only improving the input physics and their numerical modelling, but developing software that takes advantage of the latest computational architectures.

Meanwhile, another technologically driven revolution is already underway: data science. Astronomy, with its big data sets from both simulations and telescopes, has jumped on the data science bandwagon as much as any other area of science. The data rates from upcoming telescope facilities such as the Square Kilometre Array radio telescope in South Africa and Australia are in the range of terabytes per hour, which is impossible to store long-term and thus must be processed immediately and rapidly. Similarly, cosmological simulations are entering into the realm of data science, with the largest simulations taking up petabytes of disk space to store the outputs. Surely an amazing amount of information is contained in these simulations – but given a jumble of trillions of particles in a box, how does one most efficiently extract information about galaxy evolution or cosmology? This is exactly the sort of problem that data science is designed to address.

In simulations, data science techniques such as machine learning are already being employed in a variety of ways. For example, various groups have experimented with using machine learning to 'learn' the connection between galaxies and dark matter halos in a state-of-the-art hydrodynamic simulation, and then have used the results to 'paint' galaxies onto dark matter halos in a large and (relatively) inexpensive N-body simulation. This allows us to cheaply emulate all the fancy physics in the sophisticated hydrodynamic simulation within much larger volumes at a fraction of the cost, albeit with lower accuracy.

Another clever idea, recently considered, is to use machine learning to artificially boost the resolution of low-resolution simulations. Here, a high-resolution simulation is used to train a computer how to fill in information at small scales that is washed out at low resolution. This is the equivalent of sharpening an image, such as in Hollywood films when a distant blurry satellite image is, via a few clicks from the hero's trusted sidekick, sharpened to reveal the villain's licence plate. Hollywood's version may be fictionalized, but for N-body simulations it actually works surprisingly effectively. The hope is that one could run a large suite of inexpensive low-resolution simulations with varying parameters, and then use data science to refine the image to add the small-scale details at a tiny fraction of the cost.

Another application involves using machine learning to develop emulators for galaxy formation and/or cosmology. Simulations often are used to do parameter space exploration, where certain key parameters are varied over some plausible range, and the resulting simulations are compared to observations to determine which value provides the best match to the real Universe. The trouble is, individual simulations are expensive, so running a large number of them is exceedingly costly. But what if we can just run a few sparsely sampled simulations, and use machine learning to train the computer to fill the gaps? As a simple example, one might run three simulations of varied parameters – say, between 0.5, 1.0 and 1.5 – and then use machine learning to predict what the simulation would look like if it were done with a value between those figures. This is like a linear or spline interpolation, but in effect the machine learns the optimal sampling approach. This would enable much finer parameter variations, without having to tediously run each variant. This is the idea of the approach by CAMELS, a group of international research teams led by the Simons Foundation's Center for Computational Astrophysics. Already, machine learning has proved very successful in emulating parameter variations, as well as yielding intuition about how such parameter variations are reflected in galaxies and cosmology. With revolutions in artificial intelligence promised just around the corner, astronomers are sure to incorporate such approaches in the next generations of simulations.

With accelerating progress on so many fronts, what might be the state of cosmological simulations in the year 2100? It's fun to speculate, if only to allow future generations to look back and laugh at how wrong I was. Without further ado, here are my fearless predictions for the state of the art in simulating the cosmos at the turn of the next century:

We will have full galaxy formation simulations that
cover the volume of the entire Observable Universe.
This will be enabled by quantum computing
developed for the gaming industry being perverted
to calculate the laws of gravity and hydrodynamics.

Machine learning will take those simulations
and increase their resolution to hyperfine levels,
enabling all detailed galaxy properties to be predicted.

The resulting simulations will no longer output
particles, but rather radiation transport of the light
output by stars and black holes will be computed
on the fly, so that every pixel in the simulation will
be represented by a full electromagnetic emission
or absorption spectrum.

This will enable entire night sky images to be
created with better resolution than any telescope
can achieve, depicted at any cosmic epoch,
at any wavelength.

Artificial intelligence will be used to directly
compare such predicted images with real telescope
images, allowing further refinements of the models
and optimizing the constraints on baryon cycling
in an automated way.

We will have a full three-dimensional movie of
the Universe capturing all the relevant physics
required to produce galaxies like our own Milky

Way and stars like our Sun out of the primordial soup
of the Big Bang.

Planetarium shows will be mind-bogglingly awesome.

If the past is prologue, then the reality seventy years from now
will probably be far different, more mundane in some ways but
also full of fascinating and unexpected surprises. Stay tuned for
future developments . . . the adventure is just getting started!

DO WE LIVE IN A SIMULATION?

Admit it: when you picked up this book, *this* is the question you
wanted to know the answer to, isn't it?

As a cosmological simulator myself, this hits pretty close to
home. After all, if we do live in a simulation, that means that somewhere there is a Great Simulator (like me, but a lot smarter and
with a much bigger computer), who designed and ran the simulation that we call our Universe. And if that's so, then I can see
myself as being part of a chain that eventually begets some great^N-
grandchild Simulator who themselves will make their own
universe. And if *that's* so, then is the Simulator who created our
Universe just another in a long line of such Simulators? Who was
the first Simulator? Is it Simulators all the way down? The mind
boggles.

At a superficial level, there are some intriguing parallels
between cosmological simulations and reality. Simulations have
an intrinsic fuzziness owing to their softening length, while the
Universe has an intrinsic fuzziness owing to quantum mechani-
cal wave functions. Simulations have a bunch of free parameters
that are arbitrarily chosen, while the Universe has a bunch of

numbers that we call constants of nature (like the speed of light)
with seemingly arbitrary values. Simulations have a fundamental
discreteness to them manifested as machine precision beyond
which a given quantity is unknowable, while the Universe has
a vaguely analogous discreteness associated with Heisenberg's
uncertainty principle. Simulations begin at some specified time
with pre-ordained initial conditions, and if one were living inside
a simulation, that might look an awful lot like a Big Bang whose
origin is unknown and whose 'before' is meaningless.

The Oxford philosopher Nick Bostrom is often credited with
having put this simulation argument on the map in the early
2000s. Since then, there have not only been serious discussions
of its plausibility in both philosophical and scientific circles, but
there have even been tests proposed and conducted (inconclu-
sively) to identify if there is a 'discreteness' to reality that might
indicate an underlying computational architecture. Neil deGrasse
Tyson and Elon Musk, among other luminaries, have weighed in
– and I won't even say in which direction, because frankly beyond
name recognition their thoughts on this topic carry little more
weight than anyone else's. The furore has inspired many philosophers to play amateur physicists and many physicists to play
amateur philosophers, and in my experience that sort of thing
rarely ends well.

As a simulator, I'm less curious about the philosophical or
metaphysical implications than I am about the mechanics of it.
How would one even begin to design a simulation that looks like
our reality? I'm not talking about simulating something so mundane as making realistic-looking galaxies (as non-trivial as that
is); I'm talking about simulating all the way from the subatomic
level through human scales up to cosmological scales, all in a single
simulation. The amount of physics that would need to go into

this would be, well, astronomical. Surely using the same sort of approach as we do for current galaxy formation simulations, just adding layers upon layers of complexity, can't possibly be continued up to this level. Surely the Great Simulator of our Universe would have a more intelligent approach. What might that be?

At its core, a simulation requires an algorithm that takes the Universe as it exists at this instance and steps it forward in time to the next instance. But this algorithm would, I expect, have to be fundamentally different from the sort that are used for a cosmological galaxy formation simulation. In current simulations, physical laws are implemented via discretizing equations such as the law of gravity over a set of particles. The algorithm then computes this equation for all the particles, and uses the resulting forces to advance the system. But what if we had no equations to start with? What if all we had was some region of space, with no laws of physics or even any particles? How would we design an algorithm that advances this 'system', without any laws of nature as we know them? What could such an algorithm even look like?

Perhaps the most intriguing idea I have heard that is relevant to this comes from Stephen Wolfram. Wolfram's Theory of Everything claims, somewhat bombastically, that one can generate the equations of relativity and quantum mechanics from a very simple set of rules that connect points on a hypergraph. In contrast to a set of equations that describes reality as in traditional physics, Wolfram argues instead for a set of computational rules from which reality as we see it emerges. The rules operate exactly like an algorithm: the rules take a particle from one instant to the next in time. The difference from our current simulations is that the rules are purely local, that is, they don't depend on infinite-range forces like gravity, and are not discretizations of any physics equations. Wolfram has shown that even from a fairly simple set of rules, the emergent complexity is remarkable, generating all sorts of different shapes and manifolds.   '

The emergence of physical laws as we know them comes from interpreting the hypergraph in terms of our more familiar notions of space-time. In Wolfram's theory, space is a discrete set of points on a manifold, connected by rules, moving forward in time. Such space only appears smooth to us because the discreteness is on far too small a scale for us to notice. Wolfram argues that such a manifold has geodesics (that is, shortest paths), which, when represented mathematically, look like the equations of general relativity. Time then becomes nothing but the order of causal relationships in a system, and while the causality must be maintained, the progress of time can appear differently to different observers, exactly as in special relativity. A built-in limiting speed even emerges from this, which comes from the maximum speed above which causal invariance would be broken; this is then the speed of light. What we call energy is the rate that causal edges pass by an observer, and what we call rest mass is the energy for an observer whose motion is purely time-like; when these quantities are quantified on a hypergraph, their relation is $E = mc^2$.

Wolfram calls his idea the Theory of Everything because, amazingly, some principles of quantum mechanics emerge from this very same idea of causal invariance, except now applied to ensuring that certain moments of time are 'fixed' by observation. Those moments must be causally connected in a given order, but the paths between those events can be arbitrary along any number of different connecting branches within the manifold. This corresponds to the sort of indeterminate behaviour of quantum systems, with the wave function 'collapsing' (in quantum lingo) only at specific moments in time when the system is observed. Arguments over the philosophical nature of quantum mechanics and whether Schrödinger's cat is alive or dead are moot – the underlying nature of space is intrinsically 'branchial' in between frozen points in time where the system is deterministically observed. When expressed mathematically, this has a form like Heisenberg's uncertainty principle.

Wolfram's idea sounds amazing and elegant at face value, and I won't pretend to understand all the details. But it is not so much a predictive theory as it is a general framework. At its core, it requires some (presumably simple) set of rules to describe our Universe. Perhaps the first of these 'rules' was set up at the time of the Big Bang, and the complexity we see in the real Universe, and which we express via the equations of physics, is an emergent phenomenon from these rules applied over long periods of time.

The main stumbling block for this idea is that it's not at all clear what set of rules would result in the Universe that we see. Wolfram conjectures that perhaps the Universe obeys all rules for interconnecting points at once, randomly selecting one for every timestep and every point in space, but the way we choose to represent the Universe picks certain rules over others. In any case, Wolfram's theory seems more like a vague framework that could plausibly generate everything we know, under certain as yet undetermined conditions.

Nonetheless, Wolfram's idea is a fascinating one from a computational viewpoint. What if the Universe was fundamentally just a set of computational rules that generate space-time and everything in it? If Wolfram's grandiose claims are true, then the equations of physics are not the fundamental thing – the rules are. The goal of cosmology, and indeed of fundamental physics, would then be not to determine the equations describing the Universe – as has been pursued since the time of Copernicus – but rather the rules that determine the algorithm of the Universe. Perhaps if computers and algorithms had been commonplace in the early twentieth century when the revolution in physics brought about our revered equations of relativity and quantum mechanics, this sort of thinking wouldn't seem so foreign to us physicists. Perhaps we've been barking up the wrong tree this whole time.

Wolfram isn't the first outsider claiming to have an idea that will overturn all of physics. But he has a particularly interesting life story: a former star academic, Eton-, Oxford- and Caltech-educated, and the youngest ever MacArthur Prize winner, who famously turned his back on academia to start a company that produced the well-used Mathematica software, among other things. That said, if I had a dollar for every wacky theory of cosmology or physics that has shown up unsolicited in my inbox, I could retire to a tropical beach tomorrow. Wolfram's idea may simply be another one of these, only with better funding and fancier packaging.

Still, if I had to design an algorithm to simulate a universe, something akin to Wolfram's theory is a very elegant approach. It would require programming a set of timestep rules into a computer, rather than a horde of complex discretized equations. Each point in space evolves locally and independently under its own set of rules, which is trivially parallelizable and seems to be what the Universe does innately – real objects don't seem to wait for information from some faraway place in order to know how to move. The incredible complexity of the Universe is not fundamental but an emergent phenomenon, which I must say I find attractive as someone who is increasingly daunted by the disjointed sprawl of modern physics required to describe everything in our cosmos from the small to the large.

If Wolfram's Theory of Everything is even notionally correct, in my view it would dramatically increase the likelihood that we live in a simulation. I could see it being a tractable problem to programme this framework into some sort of uber-computer, even if I couldn't begin to imagine what such an uber-computer would look like, nor what rules would be appropriate. Our Universe might even be just one such simulation, with other Universes emerging from different sets of rules. We could all simply be one of a collection of simulations created as part of a parameter space exploration being run by some Great Simulator, designed to help determine the true rules of some Uber-Universe which the Great Simulator inhabits . . . which itself might just be a simulation! The mind boggles.

If our Universe is fundamentally governed by rules and not equations, perhaps all the physics we know and love today will one day be regarded as a well-intentioned but misguided historical relic, akin to phlogiston and aether. The accelerating pace of scientific progress virtually guarantees that in the not-too-distant future there will surely be another revolution in physics that ushers in a new and unfathomably bizarre worldview. It is as good a guess as any that this next revolution will emerge from the ranks of those simulating the cosmos.

SELECT BIBLIOGRAPHY

Alcock, Charles, 'The MACHO Project: Microlensing Results from
    5.7 Years of Large Magellanic Cloud Observations', *Astrophysical
    Journal*, DXLII/1 (2000), pp. 282–307
Appleby, Sarah, et al., 'The Low-Redshift Circumgalactic Medium in
    SIMBA', *Monthly Notices of the Royal Astronomical Society*, DVII/2
    (2021), pp. 2382–404
Barnes, Joshua, and Piet Hut, 'A Hierarchical *O(N log N)* Force-
    Calculation Algorithm', *Nature*, CCCXXIV/6096 (1986), pp. 446–9
Behroozi, Peter, et al., 'UNIVERSEMACHINE: The Correlation
    between Galaxy Growth and Dark Matter Halo Assembly from
    z = 0–10', *Monthly Notices of the Royal Astronomical Society*,
    CDXXXVII/3 (2019), pp. 3143–94
Boylan-Kolchin, Michael, et al., 'Resolving Cosmic Structure
    Formation with the Millennium-II Simulation', *Monthly Notices
    of the Royal Astronomical Society*, CCCXCVIII/3 (2009),
    pp. 1150–64
Colless, Matthew, et al., 'The 2dF Galaxy Redshift Survey: Spectra
    and Redshifts', *Monthly Notices of the Royal Astronomical Society*,
    CCCXXVIII/4 (2001), pp. 1039–63
Davé, Romeel, 'SIMBA: Cosmological Simulations with Black Hole
    Growth and Feedback', *Monthly Notices of the Royal Astronomical
    Society*, CDLXXXVI/2 (2019), pp. 2827–49
——, Kristian Finlator and Benjamin D. Oppenheimer, 'An Analytic
    Model for the Evolution of the Stellar, Gas and Metal Content
    of Galaxies', *Monthly Notices of the Royal Astronomical Society*,
    CDXXI/1 (2012), pp. 98–107
de Lapparent, Valerie, Margeret J. Geller and John P. Huchra,
    'A Slice of the Universe', *Astrophysical Journal Letters*, CCCII/1
    (1986), p. L1
Dekel, Avishai, and Joseph Silk, 'The Origin of Dwarf Galaxies,
    Cold Dark Matter, and Biased Galaxy Formation', *Astrophysical
    Journal*, CCCIII/1 (1986), p. 39

Doeleman, Sheperd, et al., 'Jet-Launching Structure Resolved Near the Supermassive Black Hole in M87', *Science*, CCCXXXVIII/6105 (2012), p. 355

Frenk, Carlos S., et al., 'Cold Dark Matter, the Structure of Galactic Haloes and the Origin of the Hubble Sequence', *Nature*, CCCXVII/6038 (1985), pp. 595–7

Hockney, Roger W., et al., *Computer Simulation Using Particles* (New York and Oxford, 1988)

Holbrook, Jarita C., 'Celestial Women of Africa', 28 October 2020, *arXiv*, https://arxiv.org/abs/2006.16647

Hopkins, Philip F., et al., 'Galaxies on FIRE (Feedback In Realistic Environments): Stellar Feedback Explains Cosmologically Inefficient Star Formation', *Monthly Notices of the Royal Astronomical Society*, CDXLV/1 (2014), pp. 581–603

Hoskin, Martin A., 'The Great Debate: What Really Happened', *Journal for the History of Astronomy*, XVII/1 (1978), p. 169

Hubble, Edwin P., 'Extragalactic Nebulae', *Astrophysical Journal*, LXIV/1 (1926), pp. 321–69

—, and Milton L. Humason, 'The Velocity–Distance Relation among Extra-Galactic Nebulae', *Astrophysical Journal*, LXXIV/1 (1931), p. 43

Katz, Neal, and James E. Gunn, 'Dissipational Galaxy Formation. I. Effects of Gasdynamics', *Astrophysical Journal*, CCCLXXVII/1 (1991), p. 365

—, David H. Weinberg and Lars E. Hernquist, 'Cosmological Simulations with TreeSPH', *Astrophysical Journal Supplement*, CV/1 (1996), p. 19

Kravtsov, Andrey V., and Stefano Borgani, 'Formation of Galaxy Clusters', *Annual Review of Astronomy and Astrophysics*, L/1 (2012), pp. 353–409

Li, Yin, et al., 'AI-assisted Superresolution Cosmological Simulations', *Proceedings of the National Academy of Sciences*, CXVIII/19 (2021), pp. id.e2022038118

Mac Low, Mordecai-Mark, and Andrea Ferrara, 'Starburst-Driven Mass Loss from Dwarf Galaxies: Efficiency and Metal Ejection', *Astrophysical Journal*, DXIII/1 (1999), pp. 142–55

McNamara, Brian R., and Paul E. J. Nulsen, 'Heating Hot Atmospheres with Active Galactic Nuclei', *Annual Review of Astronomy and Astrophysics*, XLV/1 (2007), pp. 117–75

Men, Hunbatz, *The 8 Calendars of the Maya: The Pleiadian Cycle and the Key to Destiny* (Rochester, VT, 2009)

Monaghan, J. J., 'Smoothed Particle Hydrodynamics', *Annual Reviews of Astronomy and Astrophysics*, XXX/1 (1992), pp. 543–74

Naab, Thorsten, and Jeremiah P. Ostriker, 'Theoretical Challenges in Galaxy Formation', *Annual Review of Astronomy and Astrophysics*, LV/1 (2017), pp. 55–109

Papovich, Casey, et al., 'ZFOURGE/ CANDELS: On the Evolution of M* Galaxy Progenitors from z = 3 to 0.5', *Astrophysical Journal*, DCCCIII/1 (2015), p. 26

Pettini, Max, et al., 'The Ultraviolet Spectrum of MS 1512-CB58: An Insight into Lyman-Break Galaxies', *Astrophysical Journal*, DXXVIII/1 (2000), pp. 96–107

Pillepich, Annalisa, et al., 'Simulating Galaxy Formation with the IllustrisTNG model', *Monthly Notices of the Royal Astronomical Society*, CDLXXIII/3 (2018), pp. 4077–106

Rees, Martin J., and Jeremiah P. Ostriker, 'Cooling, Dynamics and Fragmentation of Massive Gas Clouds: Clues to the Masses and Radii of Galaxies and Clusters', *Monthly Notices of the Royal Astronomical Society*, CLXXIX/1 (1977), pp. 541–59

Rubin, Vera C., and W. Kent Ford Jr, 'Rotation of the Andromeda Nebula from a Spectroscopic Survey of Emission Regions', *Astrophysical Journal*, CLIX/1 (1970), p. 379

Schaller, Matthieu, et al., 'SWIFT: Using Task-Based Parallelism, Fully Asynchronous Communication, and Graph Partition-Based Domain Decomposition for Strong Scaling on More than 100,000 Cores', *arXiv*, MDCVI/1 (2016), https://arxiv.org/abs/1606.02738

Schaye, Joop, et al., 'The EAGLE Project: Simulating the Evolution and Assembly of Galaxies and Their Environments', *Monthly Notices of the Royal Astronomical Society*, CDXLVI/1 (2015), pp. 521–54

Scheub, Harold, *A Dictionary of African Mythology: The Mythmaker as Storyteller* (Oxford, 2000)

Somerville, Rachel S., and Romeel Davé, 'Physical Models of Galaxy Formation in a Cosmological Framework', *Annual Reviews of Astronomy and Astrophysics*, LIII/1 (2015), pp. 51–113

Springel, Volker, and Lars Hernquist, 'Cosmological Smoothed Particle Hydrodynamics Simulations: A Hybrid Multiphase Model for Star Formation', *Monthly Notices of the Royal Astronomical Society*, CCCXXXIX/2 (2003), pp. 289–311

Springel, Volker, et al., 'Simulations of the Formation, Evolution and Clustering of Galaxies and Quasars', *Nature*, CDXXXV/7042 (2005), pp. 629–36

Tumlinson, Jason, et al., 'The COS-Halos Survey: Rationale, Design, and a Census of Circumgalactic Neutral Hydrogen', *Astrophysical Journal*, LXXVII/1 (2013), p. 59

Tumlinson, Jason, Molly S. Peeples and Jessica K. Werk, 'The Circumgalactic Medium', *Annual Review of Astronomy and Astrophysics*, LV/1 (2017), pp. 389–432

Vazquez, J. Alberto, Luis E. Padilla and Tonatiuh Matos, 'Inflationary Cosmology: From Theory to Observations', *Revista Mexicana de Física E*, XVII/1 (2020), pp. 73–91, https://arxiv.org/abs/1810.09934

Vogelsberger, Mark, et al., 'Introducing the Illustris Project: Simulating the Coevolution of Dark and Visible Matter in the Universe', *Monthly Notices of the Royal Astronomical Society*, CDXLII/2 (2014), pp. 1518–47

Walch, Stefanie, et al., 'The SILCC (SImulating the LifeCycle of molecular Clouds) Project – I. Chemical Evolution of the Supernova-Driven ISM', *Monthly Notices of the Royal Astronomical Society*, CDLIV/1 (2015), pp. 238–68

Wang, Liang, et al., 'NIHAO project - I. Reproducing the Inefficiency of Galaxy Formation across Cosmic Time with a Large Sample of Cosmological Hydrodynamical Simulations', *Monthly Notices of the Royal Astronomical Society*, CDLIV/1 (2015), pp. 83–94

Weinberg, David H., et al., 'Observational Probes of Cosmic Acceleration', *Physics Reports*, DXXX/2 (2013), pp. 87–255

White, Simon D. M., and Carlos S. Frenk, 'Galaxy Formation through Hierarchical Clustering', *Astrophysical Journal*, CCCLXXIX/1 (1991), p. 52

—, and Martin J. Rees, 'Core Condensation in Heavy Halos: A Two-Stage Theory for Galaxy Formation and Clustering', *Monthly Notices of the Royal Astronomical Society*, CLXXXIII/1 (1978), pp. 341–58

Wolfram, Stephen, 'Finally We May Have a Path to the Fundamental Theory of Physics . . . and It's Beautiful', 14 April 2020, *Stephen Wolfram Writings*, 21 April 2022, https://writings.stephenwolfram.com

York, Donald, et al., 'The Sloan Digital Sky Survey: Technical Summary', *Astronomical Journal*, CXX/3 (2000), pp. 1579–87

ACKNOWLEDGEMENTS

This book is borne from my extraordinary (and continuing) journey in the remarkable field of galaxy formation simulations. This journey would not have been possible without the guidance from leading figures during my early career such as Lars Hernquist, Joel Primack, Neal Katz, Sandra Faber, David Spergel and Jeremiah Ostriker, among others. I am indebted to my numerous colleagues around the world who have supported and challenged me, creating a thoroughly enjoyable atmosphere of collegiality and openness even while we all compete to unravel the grandest mysteries of the cosmos. I thank Jim Geach for putting forward my name to write this book, which prompted a fascinating retrospection on how this field has progressed over the past several decades, and perhaps even some reflection on my own small part therein. Finally, I am eternally grateful to my wonderful family, who continue to provide unwavering love and support throughout this thrilling ride that I call my life.

PHOTO ACKNOWLEDGEMENTS

The author and publishers wish to express their thanks to the below sources of illustrative material and/or permission to reproduce it.

S. Appleby, the SIMBA Simulation Team: 42; M. Boylan-Kolchin et al. (2009), Millennium-II Simulation Team: 20; Romeel Davé: 1, 2, 3, 4, 5, 9, 10, 11, 12, 14, 15, 16, 18, 19, 22, 23, 27, 33, 34, 36, 37, 43; EHT Collaboration: 39; ESA/Hubble and NASA: 24 right (acknowledgement: Nick Rose), 26; ESA and the Planck Collaboration: 6, 7; A. Feild/STScI, J. Tumlinson, M. Peeples and J. Werk: 35; P. Hopkins, Caltech: 29; P. Hopkins, C. A. Fauchér-Giguere, D. Keres and the FIRE team: 13 right; http://galformod. mpa-garching.mpg.de/mxxlbrowser: 13 left; http://icc.dur.ac.uk/Eagle: 13 centre; visualization by A. Klepitko using data from Walch et al. (in prep.) and Haid et al. (2019): 31; NASA/CXC/IoA/A.Fabian et al.: 41; NASA, ESA, D. Batcheldor and E. Perlman (Florida Institute of Technology), the Hubble Heritage Team (STScI/AURA), and J. Biretta, W. Sparks and F. D. Macchetto (STScI); NASA, ESA, S. Baum and C. O'Dea (RIT), R. Perley and W. Cotton (NRAO/AUI/NSF), and the Hubble Heritage Team (STScI/AURA): 40; NASA/ESA, S. Bianchi, A. Laor, M. Chiaberge, hubblesite.org; P. F. Hopkins and the FIRE simulation team: 44; NASA, ESA, D. Coe (NASA Jet Propulsion Laboratory/California Institute of Technology, and Space Telescope Science Institute), N. Benítez (Institute of Astrophysics of Andalucía, Spain), T. Broadhurst (University of the Basque Country, Spain) and H. Ford (Johns Hopkins University, USA): 21; NASA, ESA, CXC, SSC and STScI: 24 left; NASA, ESA, the Hubble Heritage Team (STScI/AURA)-ESA/Hubble Collaboration and A. Evans (University of Virginia, Charlottesville/NRAO/Stony Brook University), K. Noll (STScI) and J. Westphal (Caltech): 38; NASA, ESA, G. Illingworth, D. Magee and P. Oesch (University of California, Santa Cruz), R. Bouwens (Leiden University), and the HUDF09 Team: 25; NASA, ESA, M. Kornmesser, the CANDELS team (H. Ferguson): 28; NASA, ESA, Adam Riess (Space Telescope Science Institute, Baltimore, MD): 8; NASA, ESA, N. Smith (University of California, Berkeley) and The Hubble Heritage

Team (STSCI/AURA): 32; NASA, ESA, Y. Yang (Texas A&M University and Weizmann Institute of Science, Israel): 30 (acknowledgment: M. Mountain (AURA) and The Hubble Heritage Team (STSCI/AURA)); J. Peacock, 2dF Survey Team: 17.

INDEX

Illustration numbers are in *italics*

absorption lines 152–3
Alpher, Ralph 36–7

Baryon Acoustic Oscillations
  (BAO) 44–7
baryon cycle 149–51, 154,
  170–74, *35*
  mass balance equation 172–3
  parameters 171–2
baryonic (ordinary) matter 42,
  58–9, 100
Big Bang 24, 30, 32, 58, 61
Big Bang Nucleosynthesis 36,
  49, 116
black hole (supermassive) 116, *39*
  accretion 162
  feedback 163
  jets 158–62, *41*
box size (L) 81, 86
Brahe, Tycho 17

CANDELS survey *28*
Carina Nebula *32*
Charge-Parity (CP) violation 35
circum-galactic medium (CGM)
  151–4, *36, 40*
  hot (X-ray emitting) 160–61
  ultraviolet absorption 151–3
clusters (of galaxies) 90, 99, *21*
co-moving coordinates 65–7, *12*
Concordance (LCDM)
  cosmology 22, 92, 95, 145, 170

Copernican Principle 20, 60
Copernicus, Nicolaus 17
cosmic microwave background
  (CMB) 36–7, 42–5, 54–5, 58,
  65, 84
  fluctuations *6*
  harmonic power spectrum *7*
cosmic star formation rate
  density (SFRD) *42*
cosmic timeline *4*
cosmic web 90–94, 168, *19, 20*
critical density 48

dark energy (L) 49, 51–4, 58, 67
dark matter 38–40, 58
data science (in astrophysics)
  177–9
disc galaxy 105–7, *23, 24*
  Milky Way 106
dynamic range 71–2, 81

Einstein, Albert 26, 32, 49–50,
  102
ellipsoidal collapse 93–4, *19*
Euclid satellite 54, 166
Event Horizon Telescope (EHT)
  159
expansion *of*/*in* space *1*

feedback 127, 145–8, 164
  ejective 146
  preventive 146–7